MICROELECTRONICS

MICROELECTRONICS

CLAYTON L. HALLMARK

TAB BOOKS
Blue Ridge Summit, Pa. 17214

FIRST EDITION —SECOND PRINTING—JULY 1978

Printed in the United States
of America

Hardbound Edition: International Standard Book No. 0-8306-6794-6

Paperbound Edition: International Standard Book No. 0-8306-5794-0

Copyright © 1976 by TAB BOOKS

Library of Congress Card Number: 75-41723

Cover photo courtesy of Analog Devices, Inc.

Preface

Over the years, the design and construction of electronic circuits have changed dramatically. New devices have been discovered and old devices improved. Electronic circuits have shrunk considerably in size. And construction techniques have shifted from using discrete components to the use of plug-in integrated circuits that already contain the complete circuit.

The needs of the technician and experimenter have also changed over the years. This book is an attempt to meet that need, by gathering together a wide variety of related material in just one volume. You might call this a theory-and-practice book. The theoretical concepts of microelectronic circuits are discussed in detail, but there is much more. You'll also find practical construction and repair information, design information on linear and digital circuits, helpful data sheets, and an interesting assortment of integrated-circuit projects and experiments.

Contents

Chapter 1

Microelectronics

We use the term *microelectronics* to refer to those processes and techniques involved in the construction of circuits that are relatively small and inseparable. Active elements such as transistors and passive elements such as resistors and capacitors are closely integrated on a single wafer or block to form a functional electronic circuit.

Microelectronics includes the construction technique referred to as *integrated circuitry*; it also includes the use of thin-film techniques; and, in some cases, it includes the use of discrete microcomponents (for example, a transistor made separately and then added to an assembled unit). The term is not used to include construction using miniature (but otherwise conventional) components with interconnecting wiring between the individual components, even though some wiring may be required between assemblies in microelectronic construction.

On the other hand, *microminiaturization* is a term generally having a broader meaning. This term is used to refer to all efforts directed toward making electronic devices as small as possible. The *micromodular construction* technique, for example, is excluded from the concept of microelectronics, although it represents a good example of microminiaturization.

MICROMINIATURIZATION

The micromodule consists of tiny wafers of insulating material, on each of which is mounted an electronic element such as a diode, capacitor, transistor, or resistor. These elements, some the size of pinheads, are connected by wire threads. Micromodules represent the last practical step in the process of miniaturizing conventional electronic devices. Hundreds of thousands of such elements can be crammed into a cubic foot of space. Microelectronic devices are even smaller than micromodules.

Purpose of Miniaturization

In conventional circuits, electrical connections are the cause of numerous failures. As electronic systems become more complex, the number of parts and connections increases, and this leads to a decrease in the potential reliability of the circuitry. A characteristic of microelectronic devices, which is even more important than their small size, is the drastic reduction in the number of soldered connections required. Using a single block of semiconductor material, it is possible to fabricate complete functional subassemblies that require connections only for inputs and outputs from the unit.

The major objective of the microelectronics trend is to develop building blocks, with each block capable of performing the function of a complete circuit or group of circuits. With this capability, the block should then have one or more of the following characteristics:

1—Perform more electronic functions per unit of volume, weight, cost, power input, and power dissipated than is possible with even the most advanced components and circuit assembly techniques currently available.
2—Possess considerably higher reliability than conventional counterparts.
3—Require a minimum of maintenance as a result of improved reliability.
4—Achieve low unit cost, making throwaway maintenance practices economically practicable.
5—Perform certain new functions that are not possible with presently available components.

Size reduction alone is not always the aim of microelectronics; small size is inherent in the technology. Although the size and weight reductions are highly significant, in some applications they do not justify the development of an entirely new system of fabrication.

The Reliability Factor

Isolating trouble to a module, rather than to a specific component part, greatly reduces the number of items to be checked and thus decreases maintenance time. The requirement of increased reliability at reduced cost is inherent in this form of maintenance. By mechanizing production and improving processing techniques, the initial cost can be drastically reduced. By incorporating automatic system checkout techniques, trouble isolation can be built into the system and becomes almost completely automatic.

Circuit Classes

The conventional circuit with normal or miniature components will continue to be used. Some components, such as magnetrons and other high-power devices, do not lend themselves to drastic size reduction. Miniature components, although they do not compare with thin-film or solid-state integrated circuitry in size reduction, will also continue to be used in military applications. Unfortunately, only a limited selection of miniature components, modules, and packaged assemblies is available. It is estimated that 25—30% of all needs for electronic circuits will continue to be filled by normal or miniature components for years to come.

Discrete Components. To reduce weight and bulk, microminiature capacitors, resistors, transistors, and diodes are currently being used in conjunction with passive thin films. By micropackaging techniques, size and weight reduction is realized, but actually connecting them into usable circuits is a time-consuming and tedious operation. Such soldered connections do not realize full reliability potential.

Hybrid Circuits. Hybrid circuits represent the next logical step in fabrication. They utilize thin-film subassemblies of passive elements to which the active elements are attached. Thus the number of soldered connections is reduced and the number of wafers is also decreased. This method is a slightly improved example of microminiaturization and is often included in the definition of microelectronics.

Thin-Film Devices. Thin-film microelectronic devices consist of passive elements (currently limited to resistors, capacitors, and small inductors) that are bound on substrates of glass, ceramic, Steatite, or other suitable insulating material.

All active elements and some passive elements are added to the circuit as discrete components.

Solid-State Integrated Circuits. Solid-state circuits represent an extension of silicon transistor technology. These circuits derive their name from the fact that the circuit is formed in a single block of silicon semiconductor material. They contain both active and passive circuit elements interconnected with deposited metal patterns. The selection of circuits fabricated on the solid-state principle is at present the limiting factor in utilization of the solid-state module. Considerable development emphasis is being given to this limiting factor.

TECHNIQUES

There is a difference between microelectronics and *micropackaging.* The latter generally refers to the assembly of discrete, miniaturized components within a single enclosure, with the components being electrically connected by tiny wires and deposited films. This technique is similar to conventional methods in that separate, discrete components are used. The individual component cases or enclosures may, however, be removed to save space. The individual components are then bonded to a common substrate and interconnected. The complete subassembly is hermetically sealed in a single package. Since the individual components must be individually processed, handled, and interconnected, the reliability of the assembly is decreased and the unit cost is increased. This technique does, however, approach the size and weight reduction of microelectronic circuits.

In microelectronic technology, all the equivalent components (active and passive) are fabricated in one continuous operation on an entire block of material. The components need never be handled as individual items, and all connections except input and output are an integral part of the block. This technique maximizes reliability because of the lack of separate interconnections within the individual module. Computer-controlled automatic processing techniques can standardize construction and reduce unit cost of producing the modules, making throwaway maintenance economically practicable.

Actual production and manufacturing techniques presently in use are summarized briefly in the following paragraphs.

Deposited-Film Technique

This technique involves the fabrication of circuits that are composed of thin films of metallic or semiconductor material deposited on substrates. The individual elements of these circuits cannot be disconnected from each other.

Several methods are used to deposit the film. The most commonly used method consists of evaporating a metallic or dielectric material in a high vacuum and condensing the vapor onto a thin inactive wafer. Carefully prepared masks control the areas that are exposed for deposition of the films. By successive steps of depositing metallic or dielectric films, resistors, capacitors, and conductors can be deposited onto a single wafer. While reference here is made to the conventional names of the components, note that there are no leads between the components and that the circuit interconnections are an inherent part of the process.

Other methods used to deposit the film include the *oxidation* of certain metals, *thermal decomposition* of gaseous compounds, and *"sputtering."* In the oxidation process, the metal is first deposited onto the *substrate* (wafer), and a specified section is then oxidized to obtain special characteristics. In thermal decomposition, extremely close tolerances are required in the masking, purity of the compound, temperature, and exposure time.

Some materials have extremely high melting temperatures and are therefore almost impossible to deposit by the evaporation technique. For these materials the sputtering technique has been developed. In this process, the material is boiled under extreme heat in a high vacuum. This produces small droplets which are then deposited, or sputtered, onto the substrate wafer.

Silk-Screen Technique

In this process, a mixture of special sand and fluxes is applied to a ceramic substrate. The unit is then fired in a manner similar to the ceramic painting and glazing process. Prior to firing, interconnections are screened on, using metallic inks. External connections are welded, brazed, wire-bonded, or soldered after firing. Use of the silk-screen technique results in a film of considerably greater thickness than that of the thin-film technique.

Semiconductor Growth

This technique is the one most commonly used in the production of solid-state circuits. The process begins with a single bulk-form block of semiconductor material as the substrate. Impurities are alloyed or diffused into the material to form PN junctions. (The semiconductor material itself is used in the formation of resistors and capacitors.)

In most cases, the elements themselves and their interconnections are integral parts of the semiconductor substrate. It then becomes difficult to differentiate between the elements and the substrate. Internal interconnections are reduced by simply designing one element to begin where another ends.

Manufacturing Process. The manufacturing process consists of a number of steps that use silicon dioxide masks to control the doping of the semiconductor material. By use of computer-controlled equipment, the original block of semiconductor material is progressively masked, doped, oxidized, etched, and otherwise processed to form junctions which possess the desired characteristics. Thus, a single block of material may contain the equivalent of various combinations of transistors, diodes, resistors, capacitors, and interconnecting wires, with areas provided for connecting leads, input signals, power, and output signals.

Characteristics. The performance of solid-state circuits is limited by the capacitance of the back-biased junctions that exist between the silicon substrate and the N-type islands in which the circuit elements are diffused. This parasitic capacitance has been a limiting factor in obtaining good high-frequency response.

Resistance and capacitance values are somewhat more limited in solid-state circuits than they are in thin-film networks, and inductance is as yet unattainable. Circuit resistor tolerances of closer than 20% are difficult to attain. Temperature coefficients are relatively large for all components. Capacitors have low values of Q, are polarity restrictive, and have comparatively low breakdown-voltage values.

The absence of interconnections in completely integrated circuits leads to greatly improved reliability, simplified maintenance practices, and shorter downtime losses.

Power capabilities are limited by the drift in circuit characteristics, caused by internally generated circuit heat. Power dissipation is therefore greatly restricted.

APPLICATIONS

Microminiaturized and microelectronic circuits and components are appearing in virtually all new electronic equipment. It is expected that microcomponents and microcircuits will eventually replace about 75% of those in present use. However, there are some specific applications especially well suited for their usage.

In many equipments, the module is either the basic functional unit or the smallest maintainable item. Equipments involved in automatic data processing, logical computations, sequential switching, or aerial navigation are currently being designed with microcircuit techniques. Another area particularly well suited for miniature and microcircuits is small entertainment products such as radios and tape players.

PRINTED CIRCUITS

This type circuit appears more and more in electronic equipment, and has proven itself to be equal in operation to conventional-type construction. A few simple machine operations automatically produce circuits that formerly required a production line of workers performing the same jobs by hand.

One method of manufacturing a printed circuit is the *photoetching* process. A plastic or phenolic sheet with a thin layer of copper coating may be used. The copper coating is covered with a light-sensitive enamel. A template of the circuit that will ultimately appear on the plastic sheet is placed over the sensitized surface. The entire sheet is exposed to light. The area of the copper that is exposed reacts to the light and can then be removed by an etching process.

The exposure of the printed circuit is similar to a photographic exposure. The enamel on the unexposed circuit protects the unexposed copper from the etching bath that removes the exposed copper. After the etching bath, the enamel is removed from the printed circuit by a solvent solution. This leaves the metallic surfaces in a condition that is suitable for soldering parts and connections.

BOOK CIRCUITS

An improved type of construction, from the troubleshooter's standpoint, consists of removable subassembies called *books*. These books are readily removable and have numerous internal and external test points to facilitate troubleshooting. The books are built of easily replaceable standard parts. Most test racks have plug extensions that permit any book to be raised, making all parts accessible for checking and repairing. The book is not expendable but can be easily repaired since all parts are of conventional design.

Miniature and subminiature parts are so common in today's electronic equipment that they are now considered conventional. That's why the details of the matters just introduced are so important to know.

Chapter 2

Solid-State Devices

Since the invention of the transistor in 1948, semiconductor devices have been developed and improved at an almost unbelievable rate. Solid-state devices made from semiconductor materials offer compactness, efficiency, and versatility heretofore unattainable. These devices have invaded virtually every field of science and industry and are being used in increasing numbers for communications and data-processing applications.

As its name implies, a *semi*conductor material is classed neither as a good conductor nor as a good insulator. Although it bears physical likeness to both, its electrical characteristics are unique. In the span of just a few years, its development has contributed much to the demands of the ever-advancing electronic world.

SOLID-STATE PHYSICS

Before the operation of a transistor can be understood, the semiconducting material must first be examined. A semiconductor is neither a good conductor nor a good insulator. Germanium and silicon, which are used in most semiconductors, are substances that fall into this category.

To understand the theory of the transistor, it will be helpful to consider briefly the structures of atoms and crystals. All matter is composed of one or more elements, and each element is composed of atoms. The atom is composed of the smaller units of matter, called *electrons*, *protons*, and *neutrons*. Electrons possess a *negative charge*. Protons possess an equivalent *positive charge*, although protons are more than 1800

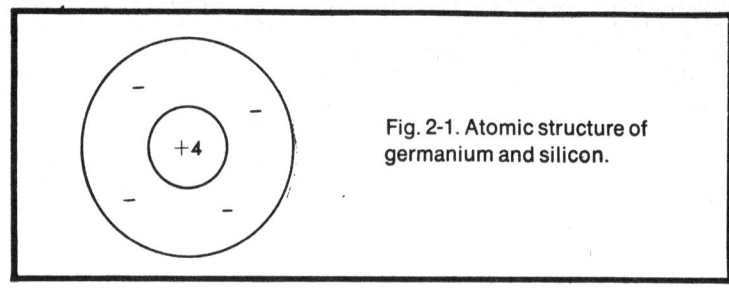

Fig. 2-1. Atomic structure of germanium and silicon.

times heavier than electrons. The neutron possesses *no charge* and has almost the same mass as the proton.

The atom, which is the smallest subdivision of an element that possesses all the properties of that element, is composed of a *nucleus*, or core, of protons and neutrons, around which the lighter electrons revolve. In the normal atom, there are as many electrons (negative charges) outside the nucleus as there are protons (positive charges) within the nucleus. Thus, the normal atom is balanced, or electrically neutral.

Atoms of one chemical element differ from those of another element only in the number of electrons, protons, and neutrons in their structure. With reference to transistors, we will consider only a few elements. These elements are germanium, silicon, antimony, arsenic, aluminum, gallium, indium, thallium, boron, and phosphorus.

The study of atomic structure has shown that a large portion of the electrons around the nucleus are tightly bound to it and do not enter into chemical reactions or transistor physics. The nucleus and the tightly bound electrons comprise an inert core having a net positive charge. Around this core the less tightly bound electrons revolve.

In transistor physics, you are concerned with the net charge on the core and with the electrons surrounding the nucleus. For example each atom of germanium has 32 protons in its nucleus and 28 tightly bound electrons around it; thus the atoms of

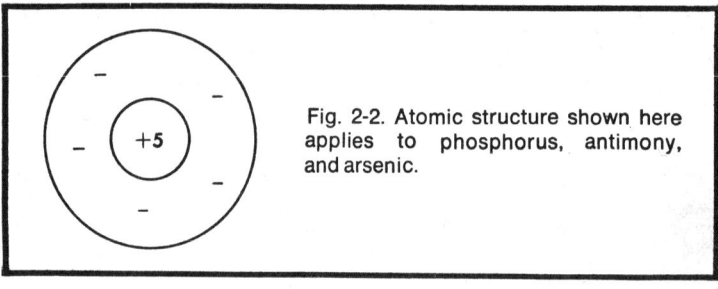

Fig. 2-2. Atomic structure shown here applies to phosphorus, antimony, and arsenic.

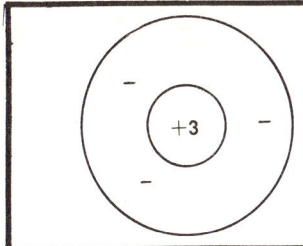

Fig. 2-3. Atomic structure of boron, aluminum, gallium, or thallium (acceptors).

germanium can be represented by a core having a net charge of +4 that is surrounded by four electrons. Each atom of silicon, which has 14 protons in its nucleus and 10 tightly bound electrons around it, is represented in the same manner as an atom of germanium, as shown in Fig. 2-1.

Pure silicon (or germanium) cannot be used in producing transistors, because silicon only acquires the properties of rectification and amplification of current through the presence of impurities in the crystals. One type of impurity is known as *donor* and another type is called *acceptor*. They have these names because of the manner in which they affect the electron movement within the transistor.

Phosphorus, antimony, and arsenic become donors when they join the crystal structure of silicon. The net charge on the cores of phosphorus, antimony, and arsenic atoms is +5. Each atom has five electrons surrounding its core, as shown in Fig. 2-2.

Boron, aluminum, gallium, indium, and thallium become acceptors when they join the crystal structure of silicon. The net charge on the cores of their atoms is +3, and each of their atoms has three electrons surrounding its core. The atomic structure of these acceptors is pictured in Fig. 2-3.

Figure 2-4 compares the atomic conditions existing in a pure silicon crystal with those existing in crystals containing

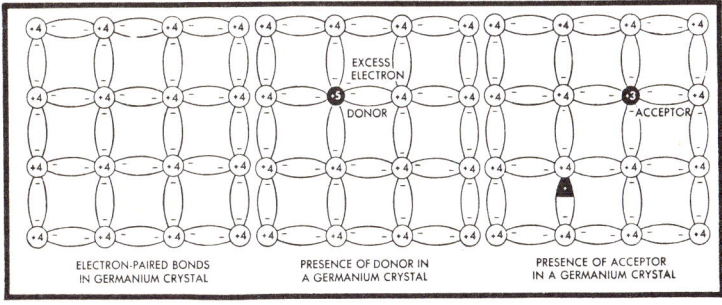

Fig. 2-4. Atomic conditions in silicon crystals.

donor and acceptor impurities. In the pure crystal, each atom has four neighbors which are equidistant from each other. Between the cores of the atoms and each of their neighbors are two electrons. These paired electrons form *electron pair bonds*, which come into existence when two or more atoms approach each other.

Since the electrons are in constant motion around the core, these electron pair bonds are formed only when the movement of an electron from one atom becomes coordinated with the movement of an electron from another atom. This coordination tends to attract the cores toward each other, but the cores' positive charges repel each other until they attain a perfect balance of attraction and repulsion. The atoms are then said to be in a condition of equilibrium.

Pure crystals cannot be used as transistors. This condition exists because the atoms of the crystals would be in a state of equilibrium. In that state, they constitute good insulators (with a dielectric constant of approximately 16), but you cannot make a transistor out of an insulator.

Effects of Donor Atoms on Silicon Crystals

When a donor atom (e.g., phosphorus) joins the crystal structure of silicon, the insulating property is lost. The donor atom must lose one of the five electrons surrounding its core, because only four of its electrons can form electron pair bonds with the electrons of its neighboring silicon atoms. The excess electron is free to move through the relatively wide spaces between the cores. It moves through the crystal as though it were in a vacuum.

If a battery were placed across the crystal, the electron would move toward its positive terminal and enter the battery at that point. Simultaneously an electron would leave the negative terminal of the battery and enter the other side of the crystal. A continuous electron flow through the crystal would take place, but the cores of the silicon and donor atoms would remain undisturbed.

Silicon crystals containing donor impurities are known as *N-type* silicon. The designation N results from the fact that conduction through the crystal is mainly a conduction of *negative charges* in the form of excess electrons from the donor atoms. It is this action that led to the term *donor* being applied to the element that, when joined to the crystal structure of silicon, gives off an excess electron.

Effects of Acceptor Atoms on Silicon Crystals

When an acceptor atom (e.g., boron) joins the crystal structure of silicon, it does so by accepting an electron from one of its neighboring silicon atoms. Atoms of acceptors have three electrons surrounding each of their cores, while the cores of silicon atoms have four electrons surrounding them. When this new electron pair bond is formed, a *hole* is created in another electron pair bond. This hole possesses the equivalent charge of an electron, but it is a *positive charge*.

Experiments have shown that the hole is free to move throughout the crystal structure as if it were an actual particle. Conduction of current can take place through the silicon containing acceptor impurities, just as through silicon containing donors.

The conduction process with acceptors in the crystal is somewhat different. If a battery were placed across this crystal, the hole would be attracted toward the negative terminal of the battery and an electron from this terminal would enter the crystal and fill the hole. Simultaneously, an electron from one of the electron pair bonds in the crystal near the positive terminal of the battery would separate from its bond and enter the battery, creating another hole in the crystal. This action would repeat and maintain a continuous flow of current through the crystal.

Silicon (or germanium) crystals containing acceptor impurities are known as P-*type* silicon. The term P is derived from the fact that the conduction through such crystals is mainly a conduction of *positive charges* (*holes*).

A *hole* can be defined as an incomplete group of electrons whose general properties are similar to those of an electron, except that it carries a positive charge instead of a negative one.

An *acceptor* is defined as an element which, when it joins the crystal structure of silicon, produces a hole, or excess positive charge, within the crystal.

JUNCTION TRANSISTORS

Doped silicon crystals are good conductors and can conduct current equally well in either direction. Rectification with a silicon crystal occurs if P-type silicon and N-type silicon are placed side by side. The plane at which the two types of silicon meet is called a PN *junction*. The action which occurs at such *junctions* produces the basic phenomenon of transistor operation.

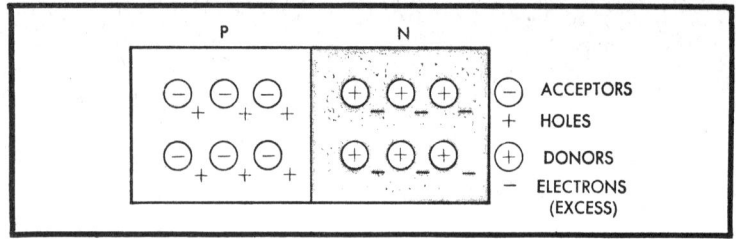

Fig. 2-5. Equilibrium condition in PN junction.

Figure 2-5 illustrates a PN junction in a state of equilibrium. Notice that the holes concentrate to the right of the P-type silicon atoms and the excess electrons concentrate to the right of the N-type silicon atoms. This phenomenon results from the distribution of electrostatic potential or *voltage* produced by the acceptor and donor atoms, as illustrated in Fig. 2-6.

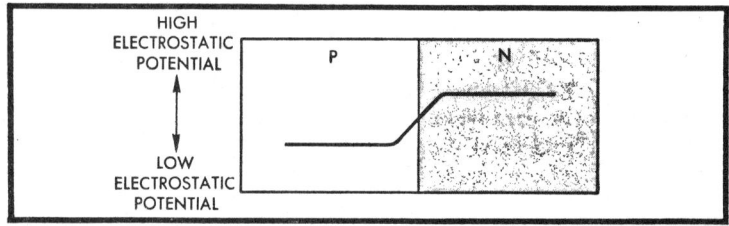

Fig. 2-6. Electrostatic potential.

The electrons remain in the region of highest electrostatic voltage, and the holes remain in the region of lowest electrostatic voltage. When an electron is in the region of highest electrostatic voltage, its potential energy is at a minimum, as shown in Fig. 2-7. Since *potential energy* means ability to do work, the electron cannot move to do work *after* it reaches the point of highest electrostatic voltage because it is in a region of lowest potential energy.

The same principle can be applied with reference to holes. When the hole is in a region of lowest electrostatic voltage (that is, low negative potential), its potential energy is at a minimum,

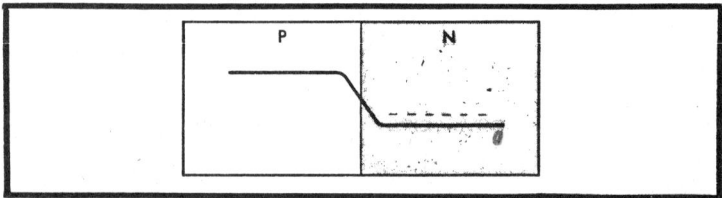

Fig. 2-7. Potential energy of electrons.

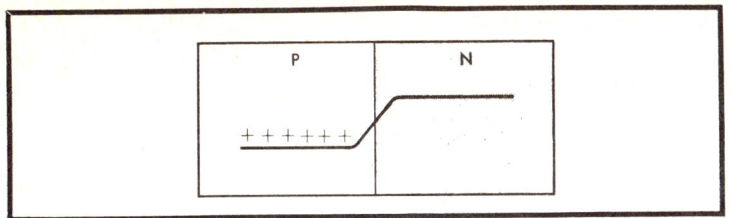

Fig. 2-8. Potential energy of holes.

as illustrated in Fig. 2-8. (Potential energy diagrams may also be used to illustrate transistor behavior.)

Holes and electrons flow into regions of low potential energy only. A low-potential-energy region for an electron is a high-potential-energy region for a hole, and vice versa.

PN Junction With Reverse Bias

If you connected a battery across a PN semiconductor junction as shown in Fig. 2-9, there would be no conduction of current through the crystal. Such a connection is called *reverse bias* and occurs when the positive terminal of the battery is connected to the N-type silicon and the negative terminal is connected to the P-type silicon.

The positive terminal of the battery attracts the electrons and causes them to concentrate farther to the right than when the junction is in a condition of equilibrium. Similarly, the negative terminal of the battery attracts the holes and causes them to concentrate farther to the left. As a result of these attractions, there is no flow of electrons to the left and no flow of holes to the right. The difference in electrostatic potential between the two types of silicon has been increased as shown in Fig. 2-10.

The potential energy "hill" for the holes has been increased. They will not flow up the steep "hill." This condition is shown in Fig. 2-11. The same condition applies to the potential

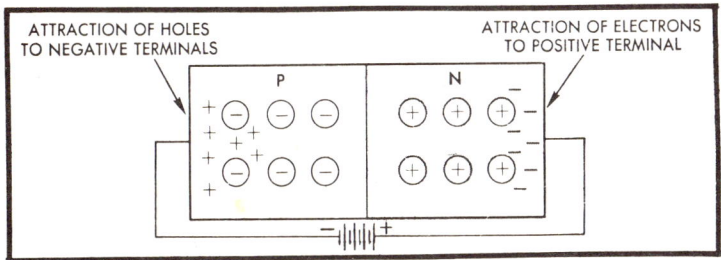

Fig. 2-9. A PN junction with reverse bias.

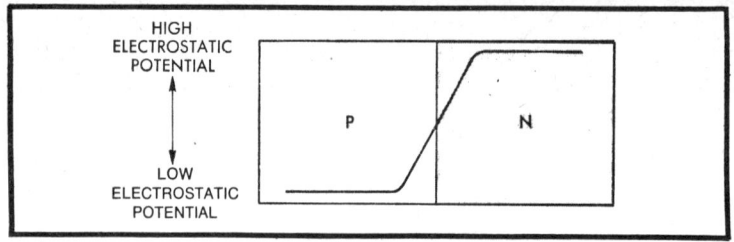

Fig. 2-10. Electrostatic potential increased in PN junction with reverse bias.

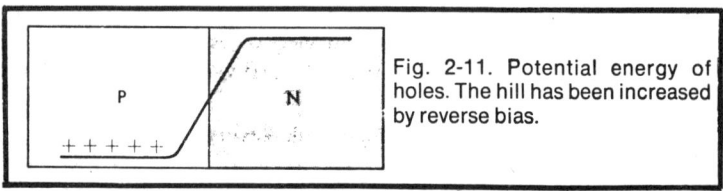

Fig. 2-11. Potential energy of holes. The hill has been increased by reverse bias.

energy "hill" for the electrons and prevents their flow, as shown in Fig. 2-12.

PN Junction With Forward Bias

If you connect a battery across a PN silicon crystal with the positive terminal connected to the P-type silicon and the negative terminal to the N-type silicon, current will flow in proportion to the applied voltage. This type of connection is known as *forward bias.*

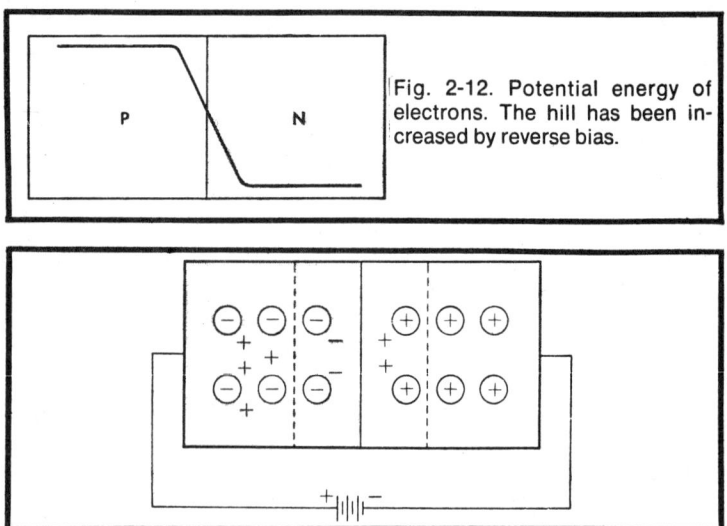

Fig. 2-12. Potential energy of electrons. The hill has been increased by reverse bias.

Fig. 2-13. A PN junction with forward bias.

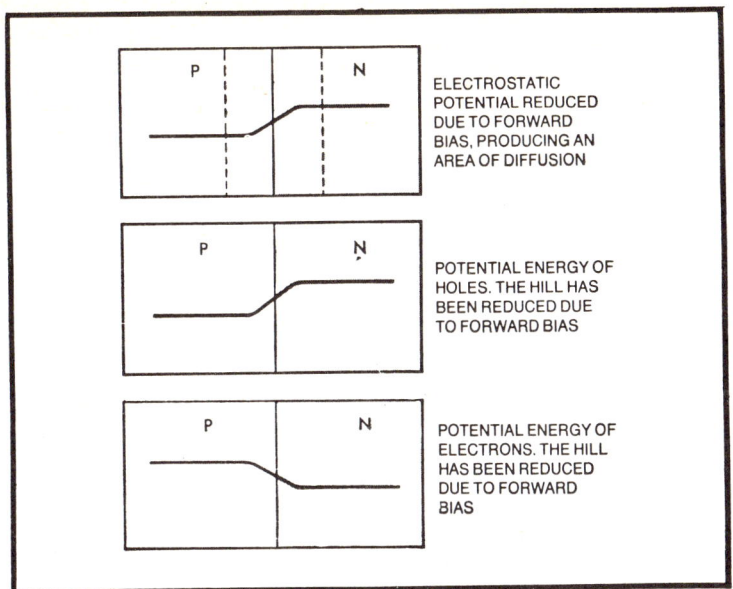

Fig. 2-14. Potential energy conditions existing under application of forward bias.

As shown in Fig. 2-13, the positive terminal of the battery repels the holes and causes them to move toward the N-type silicon. Some of the holes enter the N area. Similarly, the negative terminal of the battery repels the electrons and causes them to move toward the P-type silicon. In this case, some electrons enter the P area.

Electrons and holes combine in a small area of diffusion on either side of the PN junction (between the dotted lines in the diagram). For each hole in the P region that combines with an electron from the N region, an electron from an electron pair bond in the crystal near the positive terminal of the battery enters the battery at the positive terminal. This action creates a new hole, which moves toward the N-type silicon. For each electron that combines with a hole in the N-type silicon, an electron enters the crystal from the negative terminal of the battery. The current flow in the P region is mainly a flow of holes, while that in the N region is mainly a flow of electrons. The potential energy conditions existing under the application of the forward bias shown in Fig. 2-14.

The currents that flow through the crystal are shown in Fig. 2-15. The total current, I, is constant; current by holes, I_P, is shown by the solid line; and current by electrons, I_N, is shown by the dashed line.

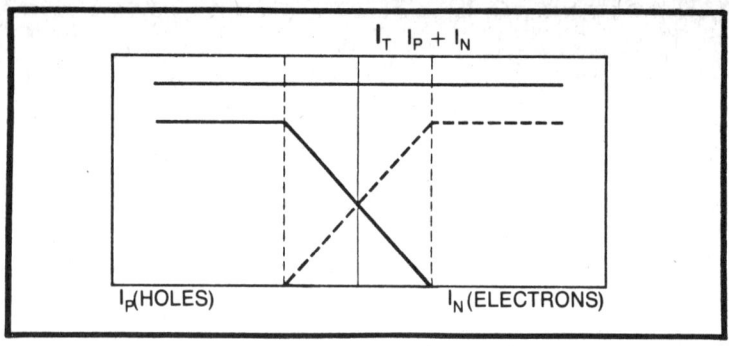

Fig. 2-15. Current flow through crystal under application of forward bias.

From this theory, you can understand more fully the process of rectification in silicon crystals. You can better understand how the crystal acts as a low resistance when forward bias is applied and how it acts as a high resistance when reverse bias is applied. These phenomena constitute the basic principles of transistor operation.

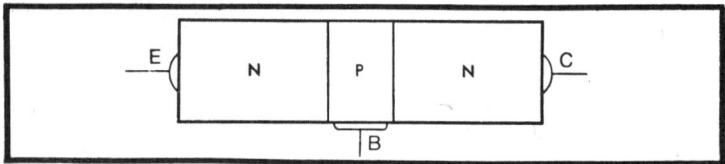

Fig. 2-16. An NPN junction transistor.

NPN AND PNP JUNCTION TRANSISTORS

An NPN junction transistor is pictured in Fig. 2-16. In this type large surface contacts (low resistance) are made to each semiconductor strip. The N-type silicon shown at the left is called the emitter, E. The N-type strip shown at the right is called the collector, C. The P-type section in the center is called the base, B.

Figure 2-17 shows the distribution of the donors, acceptors, holes, and electrons in an NPN junction transistor under equilibrium conditions (no external voltages applied). The

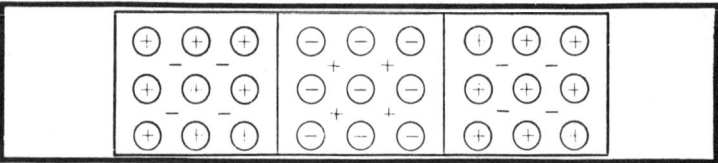

Fig. 2-17. An NPN junction transistor in state of equilibrium.

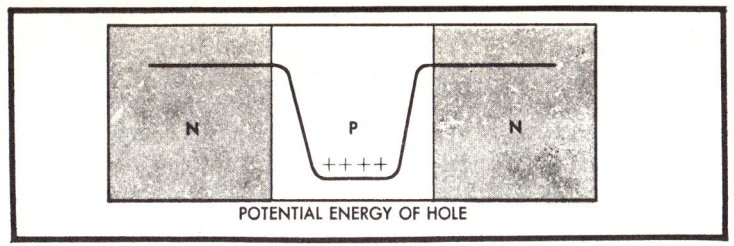

Fig. 2-18. Potential energy of holes in NPN junction transistor.

potential energy of the holes in this type of transistor is represented in Fig. 2-18. Figure 2-19 shows the potential energy region of the electrons.

The holes in Fig. 2-18 are concentrated in the region of lowest potential energy for them. They cannot "climb" the potential energy "hills" to the right or left. The electrons in Fig. 2-19 are also concentrated in the region of lowest potential

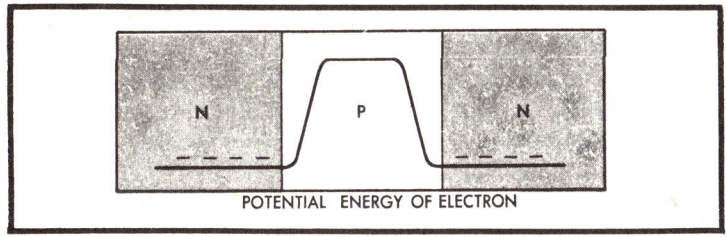

Fig. 2-19. Potential energy of electrons in NPN junction transistor.

energy for them and cannot "climb" the potential energy "hills" to enter the P-type silicon. Thus, no current flows.

In practical applications the NPN junction transistor normally is biased in the manner shown in Fig. 2-20. The PN junction between the emitter (E) and the base (B) is biased in the forward direction. The PN junction between the collector (C) and the base is biased in the reverse direction.

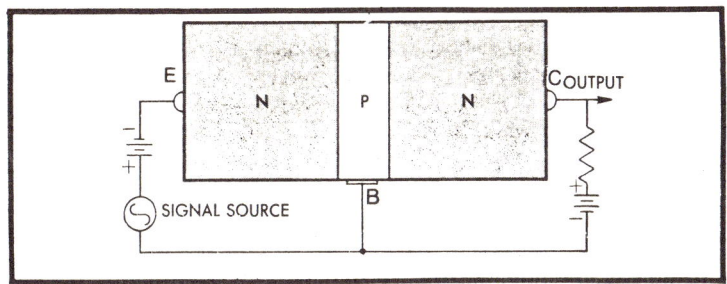

Fig. 2-20. An operating circuit with NPN junction transistor.

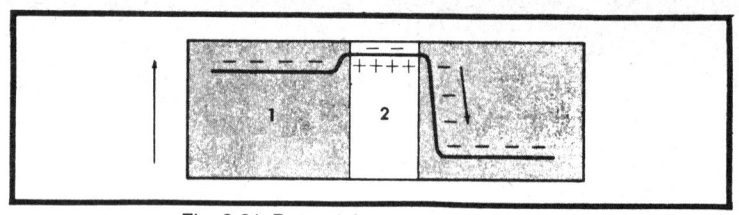

Fig. 2-21. Potential energy of an electron.

Figure 2-21 shows the potential energy condition for electrons with no signal applied. In NPN transistors the electrons are the major current components. Potential energy diagrams for holes are not shown.

The forward bias applied between the emitter and base reduces the potential energy "hill" at the left PN junction. Some electrons "climb" this "hill" and enter the P-type silicon. Since the base strip is relatively thin, most of the electrons which enter it do not combine with holes. Instead they pass through the strip and readily "go down" the potential energy "slope" at the right PN junction, as shown by the arrow in Fig. 2-22.

The steep potential energy "slope" that permits easy entrance of the electrons from the base strip to the N-type silicon of the collector is produced by the application of reverse bias between the collector and the base.

When a signal is applied that opposes the forward bias on the emitter (makes base more negative with respect to emitter), the potential energy "hill" between the emitter and base is increased, and fewer electrons "climb" the "hill" to enter the P-type silicon. This condition is illustrated in Fig. 2-23. Most of the electrons that do enter the P-type silicon still do not recombine with holes but "fall" into the collector region of low potential energy.

When the polarity of the applied signal aids the forward bias (makes base more positive with respect to emitter), the potential energy "hill" between the emitter and the base is decreased, and more electrons flow into the P-type region. This condition is demonstrated in Fig. 2-22.

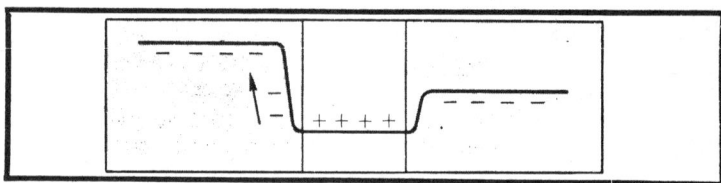

Fig. 2-22. Signal applied to PNP transistor, making base more positive with respect to emitter.

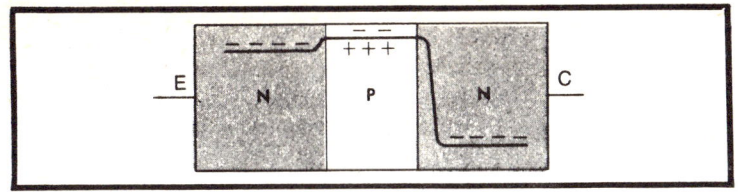

Fig. 2-23. Signal applied to NPN transistor, making base more negative with respect to emitter.

You can compare the operation of the NPN junction transistor to the operation of the triode vacuum tube. The emitter is equivalent to the cathode, the base to the grid, and the collector to the plate. Practically all of the electrons which emerge from the emitter (cathode) go to the collector (plate). The base (grid) current is minute and consists only of a small number of electrons which represent the recombination number of holes and electrons in the P-type region.

Two types of current are present in the vacuum tube, and two types of current are involved in the transistor. The thermionic electrons (those emitted from the cathode) of the tube are equivalent to the excess electrons from the emitter.

The minute conduction current of electrons (those which flow in and out of the grid to control plate current) is equivalent to the small current of holes in the base region (which varies to control collector current).

PNP Junction Transistors

The PNP junction transistor consists of a narrow strip of N-type silicon between two relatively long strips of P-type silicon. Large surface contacts (low resistance) are made to each strip. The contacts are designated B, C, and E, representing the base, collector, and emitter, as in the case of the NPN transistor. This is illustrated in Fig. 2-24.

The operation of the PNP junction transistor is analogous to that of the NPN, except that the hole constitutes the main current

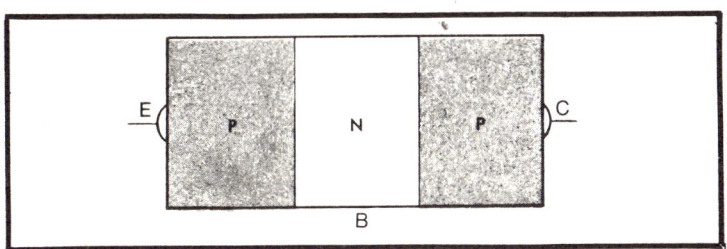

Fig. 2-24. A PNP junction transistor.

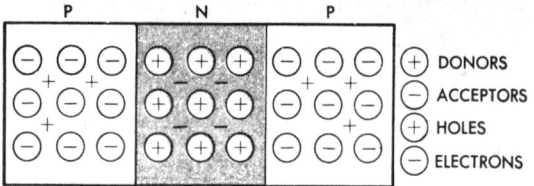

Fig. 2-25. A PNP junction transistor in state of equilibrium.

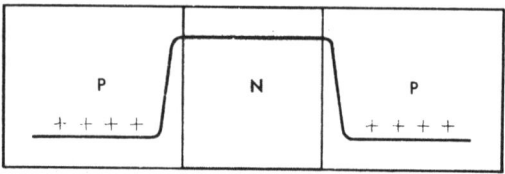

Fig. 2-26. Potential energy of holes in PNP transistor.

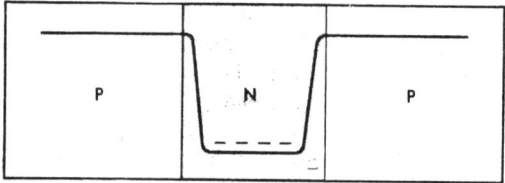

Fig. 2-27. Potential energy of electrons in PNP transistor.

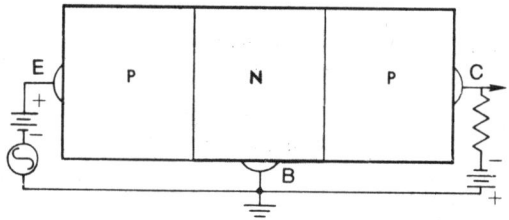

Fig. 2-28. Operating circuit with PNP junction transistor.

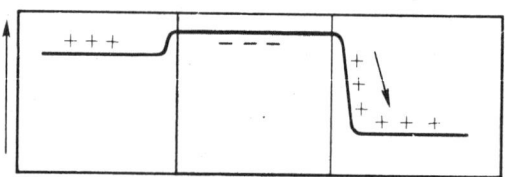

Fig. 2-29. A PNP transistor (no signal).

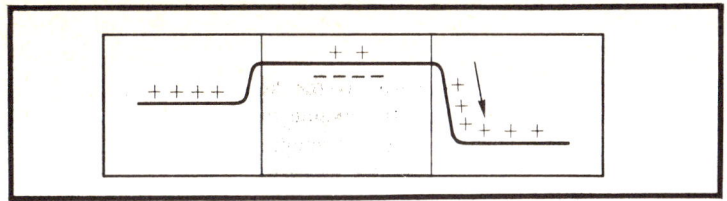

Fig. 2-30. A PNP transistor with signal applied, making base more positive with respect to emitter (hill increases).

component instead of the electron. Pictured in Fig. 2-25 is the PNP transistor in a state of equilibrium. The excess electron is the key factor in the operation of the PNP junction transistor, but the hole is the main current component. Basic conditions determining the potential energy of the holes and electrons are shown in Figs. 2-26 and 2-27. Figures 2-28 through 2-31 illustrate the conditions of operation for the PNP junction transistor in the same manner that the earlier figures illustrated the conditions of the NPN transistor.

To bias the PN junction between the emitter and the base of the PNP transistor in the forward direction, the emitter is made positive with respect to the base. To bias the collector in the reverse direction, the collector is made negative with respect to the base.

The emitter, base, and collector connections to the crystal are low-resistance (large-area) connections in a junction transistor.

MANUFACTURING PROCESSES

The raw material of which transistors are made is *trichlorosilane*—a liquid rich in silicon, extracted from sand. This chemical compound is refined into pure silicon in the form of *polycrystalline* silicon. *Polycrystalline* means *many crystals*, and even a minute piece of this kind of silicon contains many crystals. But transistors and ICs are *monolithic*, which means *of one stone*. Hence, the next stage of the crystal-manufacturing

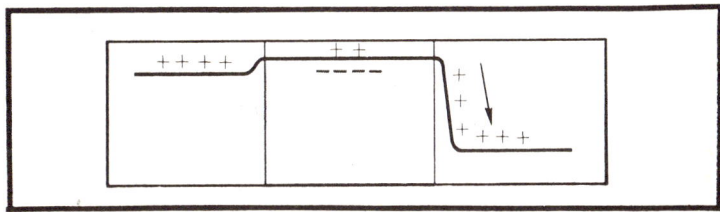

Fig. 2-31. A PNP transistor with signal applied, making base more negative with respect to emitter (hill decreases).

process is to convert a quantity of polycrystalline silicon into a single, continuous crystal. The term for this stage is *crystal growing*. What we have at this point is nothing basically new; crystals have always existed in nature. It is the process of *doping* that gives us transistors, ICs, and other electronic devices.

Grown Junctions

To produce silicon in which all of the atoms line up in exactly the right pattern, a *crystal lattice*, we begin by melting a quantity of polycrystalline silicon in a crucible about the size of a cup. Crystal growing is essentially a reproductive process; we take a little piece of monocrystalline silicon about the size of a pencil eraser and reproduce its crystal structure many times over to obtain a piece about the size of a carrot.

In this process, we lower the small piece of silicon, the *seed*, into the crucible of molten silicon until the seed just touches the surface. The molton silicon begins to crystallize on the cooler seed. The remarkable and useful thing about this process is that the new crystalline formation has the same crystalline lattice structure as the seed; and as it continues to grow, the structure remains monocrystalline. We slowly and continually pull the seed away from the melt, while rotating the seed and crucible in opposite directions. The result is a single cylindrical crystal, large enough to make millions of tiny chips for transistors, diodes, or ICs.

Next, we saw the crystal cylinder into tiny chips, all of which have just the right pattern to produce a semiconductor. To produce a useful semiconductor device, we must dope the chips in the right places and produce junctions. The three ways of producing junctions are: *growing*, *alloying*, and *diffusion*. Grown junctions, in turn, are produced by two main methods, as explained next.

Molten Method. In this method, P and N pellets are sequentially dropped into the molten silicon as the grown junction is withdrawn, forming P and N layers. This method is mainly of historical interest. It is not much used anymore, but was the original way of making germanium and silicon transistors.

Epitaxial Method. *Epitaxial* means *arrange upon* or *deposit*. In this method, a monocrystalline wafer is heated in a chamber, and silicon chloride gas carrying a dopant is admitted to the chamber. Doped silicon from the gas forms a layer on the wafer. If, for example, the wafer were N material, the gas would

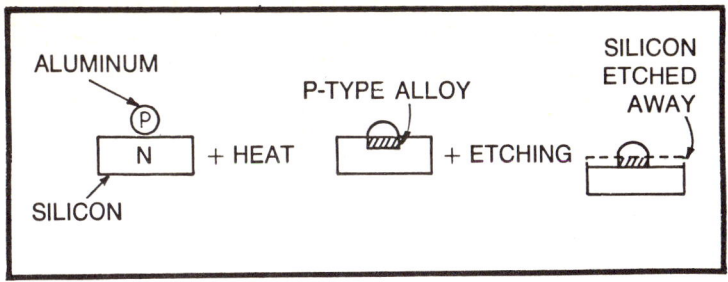

Fig. 2-32. Alloyed-junction process.

carry boron, a P-type dopant. If a transistor is to be made, the deposited region is the base. The emitter would be formed by *diffusion*, which will be discussed in a moment.

Alloyed Junctions

This process was once used to produce germanium transistors and diodes. It still is used to produce silicon alloy diodes.

A silicon crystal, grown as explained above, is cut into slices. A graphite disc, the same size as the silicon disc but with many small holes in it, is placed on top of the silicon. A tiny ball of aluminum, a P-type impurity, is placed in each hole. This assembly is placed in an oven and heated until the aluminum melts. When the aluminum melts, it combines, or in this case *alloys*, with the silicon underneath it (Fig. 2-32). When the assembly is removed from the oven, the aluminum cools and hardens, leaving many N-type aluminum bumps on top, a P-type aluminum—silicon alloy in the middle, and the original N-type silicon on the bottom.

To remove the short that would otherwise exist between each N-type aluminum bump and the N-type wafer (substrate), the entire chip is immersed in acid having an affinity for silicon but not for aluminum. Thus, a layer of silicon is washed away except where the chip is protected by the aluminum, leaving the P-type alloy regions exposed.

To make transistors, the chip can be flipped over and a P-type collector region alloyed opposite each of the P regions already formed. In any case, each chip is sawed into as many diodes or transistors as there are alloyed pedestals on one side.

Diffused Junction

How diffusion works is depicted by Fig. 2-33. We start with a wafer of monocrystalline silicon obtained as explained above.

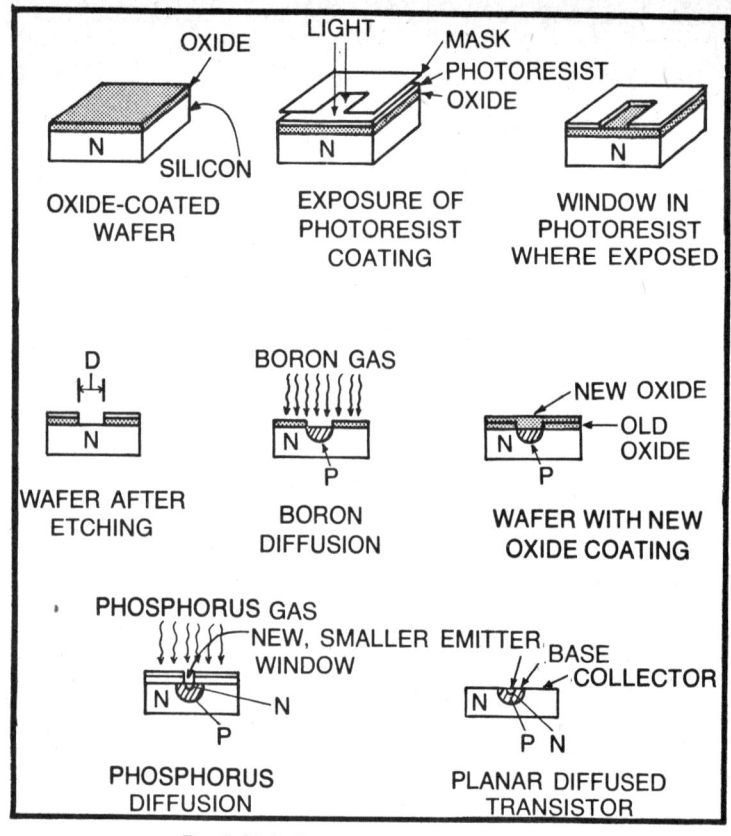

Fig. 2-33. Diffused-junction process.

On this silicon substrate, or underlayer, we obtain a protective insulating layer by means of oxidation. Next, we make a large number of holes, or *windows*, in the silicon layer by a photoetching process. Figure 2-33 illustrates this process for a single window. Actually, window forming and the whole diffusion process take place many times over, simultaneously. Therein lies the uniformity and economy realizable by the diffusion method.

To make the windows, we coat the substrate with a layer of *photoresist* and place a many-holed *mask* over it. Photoresist has the property of undergoing chemical changes when exposed to light, which makes it susceptible to etching. The holes in the mask are made where we wish to produce windows in the silicon oxide layer. Next we shine a light on the mask and expose the photoresist at selected places, as determined by the pattern of

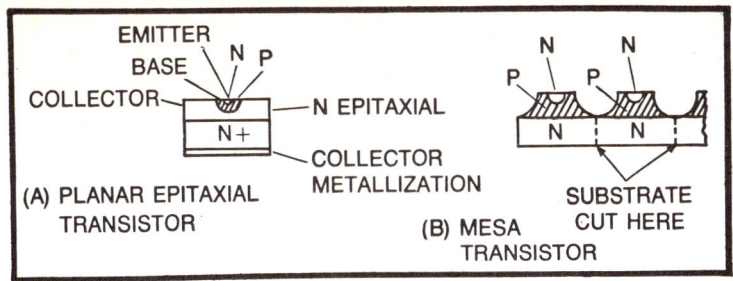

Fig. 2-34. (A) Planar epitaxial construction. (B) Mesa construction.

holes in the mask. Then we wash the slice, clearing away the photoresist in the selected places. We now wash the slice in a chemical etchant, cutting windows in the oxide layer at the selected places where the protective photoresist coating has been removed. The width of the windows (*D* in Fig. 2-33) is about 0.020 in. for a small-signal transistor, and about one-fourth that for an IC transistor. This completes the photoetching process.

We place the windowed silicon slice in a furnace, heat it to near the melting temperature, and admit gas into the furnace. The boron is unable to penetrate the oxide layer, but readily soaks (diffuses) into the silicon at the windows, forming a small P region at each window. Then we coat the entire slice with a new layer of silicon oxide, and etch a new, smaller window over each P region. This second photoetching process is the same as the first. After this step, we again expose the slice to a gas—this time to phosphorus, an N dopant. We thus produce an N region within each P region, giving us the three regions required for an NPN transistor, as shown in the drawing.

A thin layer of aluminum is placed over the emitter and base regions, a step called *metallization*. This produces electrical contacts so that the regions can be connected to the outside world. The collector metallization is usually gold. When the silicon slice is cut into transistors (several hundred are formed on one slice), the gold will serve as a very effective means of bonding the chips to their packages and transferring heat and collector current from the chips.

Transistors produced as just described are *planar diffused* types because, in the end result of the diffusion process, the N and P regions all appear at the surface of the slice, in a plane.

Planar Epitaxial and Mesa Transistors

Two other kinds of diffused transistors are illustrated in Fig. 2-34. To make a *planar epitaxial* transistor, we start with a heavily doped N substrate (N+). We epitaxially deposit a lightly doped N region for the collector. Then P and N regions are diffused into the collector as already described.

This process is particularly useful for making high-voltage transistors. The lightly doped collector is able to withstand relatively high voltages. The heavily doped substrate helps to provide a good low-resistance contact for the collector.

For mesa transistors (Fig. 2-34B) a single P layer is epitaxially deposited or diffused over the entire substrate. Then, N regions are diffused in the regular way. Next, valleys are etched in the P layer and the slice is cut into transistor chips. The purpose of the valleys is to prevent damage to the collector—base junctions during the sawing operation which separates the chips.

MOSFET

Details of the construction of an N-channel depletion-mode MOSFET are shown in Fig. 2-35. As can be seen, the main body is composed of a P-type substrate. (In this sense, a *substrate* is the physical material on which a circuit is fabricated. While its primary purpose is to furnish mechanical support, it may also serve a useful thermal or electrical function.)

The source and drain material is diffused into the body. In this case, it is a low resistivity N-type material indicated by N+ in the figure. Between the source and drain materials a moderate-resistivity N channel is also diffused into the body. Covering the main body of the MOSFET is a thin layer of oxide, allowing metallic contact to only the source and drain. The gate metal area is now overlaid in the area between the source and drain contacts on the oxide. There is no physical contact between these metal contact areas.

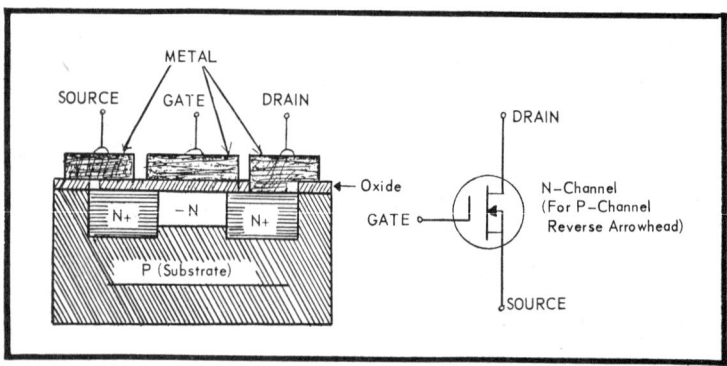

Fig. 2-35. Construction of depletion-mode MOSFET and schematic symbol.

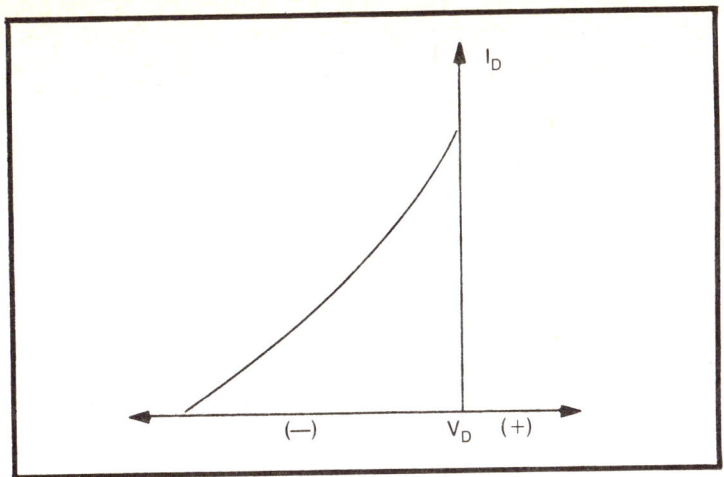

Fig. 2-36. Characteristic of depletion-mode MOSFET.

The N channel between source and drain allows current to flow when a potential is applied between the source and the drain, with a zero gate voltage. When a reverse bias voltage is applied to the gate, drain current is reduced. The greater the reverse voltage on the gate, the lower the drain current.

For the N-channel MOSFET shown in Fig. 2-35 a negative voltage applied at the gate will cause an increase in the depletion of the channel, increasing the resistance of the channel and lowering I_D. Figure 2-36 shows the effect of varying the gate voltage, V_G.

When the gate voltage on a MOSFET is zero, there will be an electric field set up across the oxide insulating material due to the drain current. The metal contact of the gate forms one side of a capacitor, the oxide material acts as the dielectric, and the N channel is the other side of the capacitor.

As the drain current flows from left to right in Fig. 2-35, the N channel will exhibit a greater positive potential. The upper plate of our capacitor, the metal of our gate, will be negative with respect to the bottom of the plate, the N channel. We have an electric field that is exhibiting a negative potential at the top of the capacitor and a positive potential at the bottom.

The increasingly positive potential, due to drain current flowing from left to right, makes the field across our capacitor stronger on the right, so that the field appears in the shape shown in Fig. 2-37. An increase in the reverse bias potential on the gate will also increase the strength of the field shown in Fig. 2-37, and the drain current will be reduced.

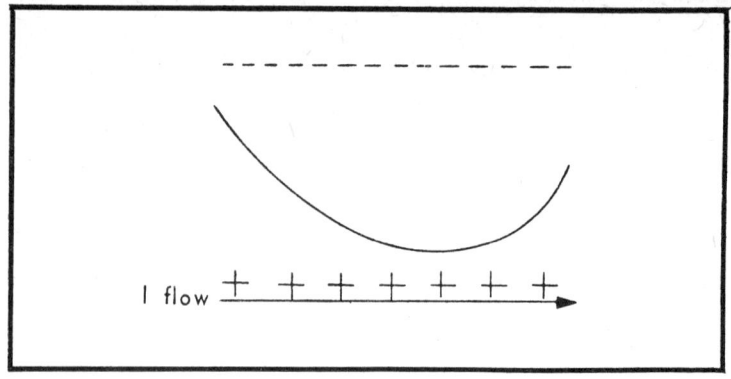

Fig. 2-37. Potential distribution.

One of the uses of the MOSFET is in electronic voltmeters. You will recall that the higher the resistance of the voltmter, the more accurate it is for making measurements in high-resistance circuits. Since the input impedance of a MOSFET is very high, it is practical to use it in the input of an electronic voltmeter. The greater the reverse bias applied to the gate of a MOSFET the lower the drain current, and, conversely, the lower the reverse bias applied to the gate, the higher the drain current. It can be seen that by connecting the drain terminal to meter circuitry, the meter will respond to changes in drain current.

INTEGRATED CIRCUITS

Another development in solid-state electronics is the integrated circuit (IC). As the term *integrated* implies, several components are contained in one package. One semiconductor chip can contain two or more transistors, several resistors and capacitors, as well as many individual diodes. How this may be accomplished is our next topic. Figure 2-38 is a pictorial and schematic of a simple IC.

As can be seen from Fig. 2-38, it is easy for one IC device to replace many separate or discrete components with a single chip by using the various properties of semiconductor elements, such as resistance, capacitance, and conductance. The circuit shown in Fig. 2-38 could by enclosed in the same size case now associated with transistors. In addition to the space-saving features of ICs, the fact that they receive almost identical processing enables them to be closely matched in characteristics. The closer such circuit elements are matched, the greater the reliability of the circuit.

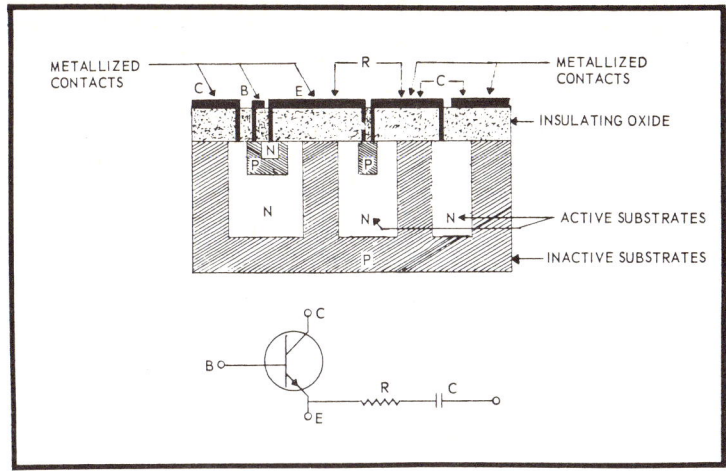

Fig. 2-38. Schematic and pictorial diagram of a simple IC device.

There is an unlimited number of circuits that could be integrated, but the one to be described at this time is the differential amplifier. A schematic symbol for this circuit is shown in Fig. 2-39, while Fig. 2-40 is the complete schematic. A differential amplifier is basically an amplifier that responds only to the difference between two input voltages or currents; it effectively suppresses equal voltages or currents.

Inspection of Fig. 2-40 shows that the signal inputs are applied to the bases of Q1 and Q2 through leads 2 and 3, respectively. These input signals are amplified by Q1 and Q2 and applied to the bases of Q3 and Q4. The outputs of emitter

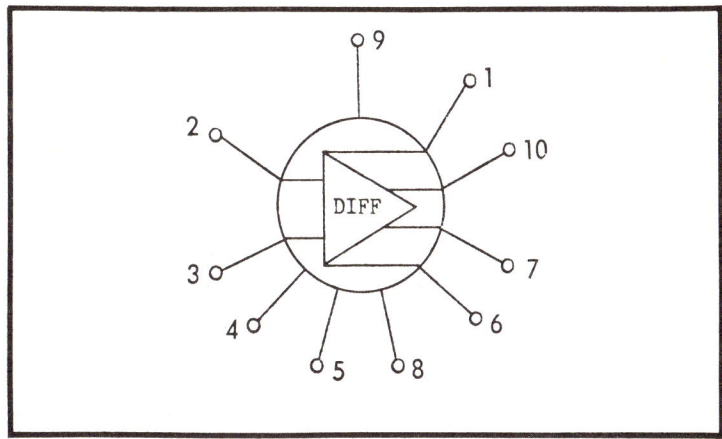

Fig. 2-39. Schematic symbol for IC differential amplifier.

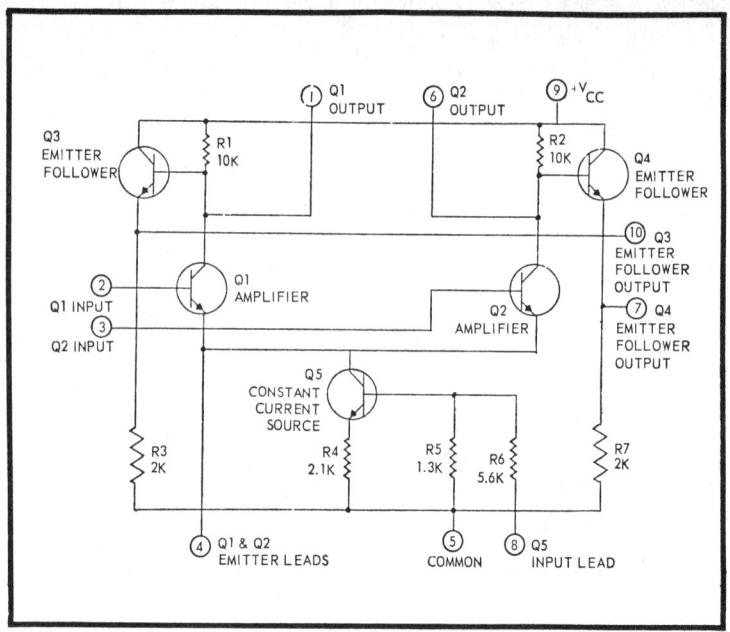

Fig. 2-40. The differential amplifier, an integrated-circuit device.

followers Q3 and Q4 are available at pins 10 and 7, respectively. The output signals of Q1 and Q2 (the collector signals) are also available at pins 1 and 6 of the device. Constant-current source Q5 determines the gain of the circuit by determining the emitter currents of Q1 and Q2. Control of the amplifier gain can be accomplished by applying a signal to the base of Q5 through pin 8 and R6. If it is desired to bypass the constant-current source, a signal may be applied directly to the emitters of Q1 and Q2 at pin 4. Lead 9 is the connecting lead for the positive V_{CC}, and lead 5 is the common connection for the device.

It should be noted that all the components shown in Fig. 2-40 are contained on the IC chip and that it might be necessary to connect external biasing or coupling components, depending upon the required operating conditions. For example, external resistors might be connected from pin 9 to pins 2 and 3 to provide bias for Q1 and Q2. This type of construction, with primary connections, increases the flexibility of the device; fewer types are required, and thus the cost is reduced. The values given for Fig. 2-40 are nominal and are not necessarily those that would be found in any particular IC.

In succeeding chapters we shall consider the differential amplifier and many other circuits in much greater detail.

Chapter 3

Microelectronic Techniques

Microelectronics is the technology of constructing electronic circuits and devices in extremely small packages by various techniques. This technology is also sometimes referred to as *microminiaturization*.

The increasing complexity of electronic systems over the past 30 years has made the evolution of microelectronics inevitable. During this period, the electron tubes in the early electronic systems have been replaced by solid-state discrete devices and integrated circuitry; and these, in turn, are giving way to medium- and large-scale integrated circuitry.

Microelectronics today encompasses thin-film, thick-film, hybrid, and integrated-circuit technology. These approaches (and combinations of them) are being applied in every branch of electronics.

The current trend of producing a number of circuits on a single chip will further increase the *packing density* of electronic circuits while reducing the size, weight, and number of connections in individual systems. Improvements in reliability and system capability are also to be expected.

Integrated circuits can be produced that combine all the elements of a complete electronic circuit on a single chip of silicon. The implications of this in the microelectronic evolution are easily demonstrated—compare a conventional J-K flip-flop circuit incorporating solid-state discrete devices and the same type of circuit employing integrated circuitry.

The conventional circuit depicted schematically in Fig. 3-1 would require approximately 40 separate discrete elements, 200 connections, 40 hermetic seals, and 300 separate processing operations, with each operation, seal, and connection

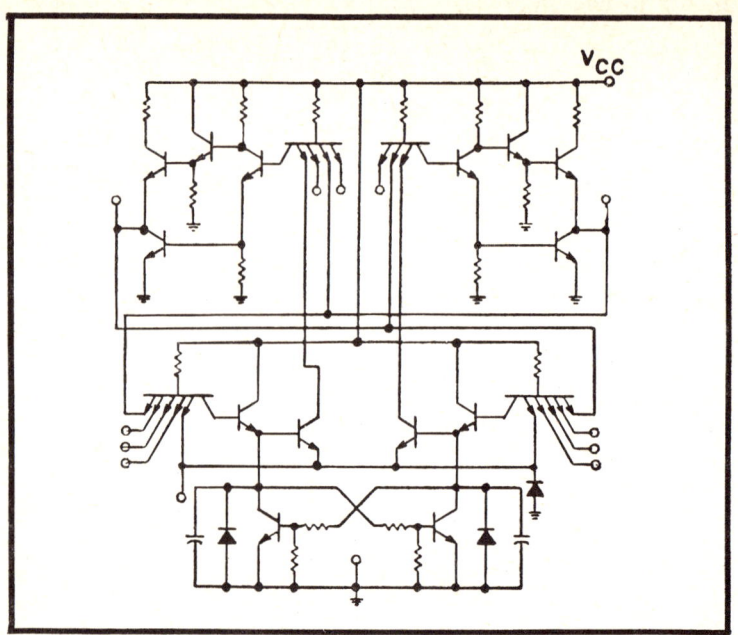

Fig. 3-1. Schematic of a J-K flip-flop.

representing a possible source of failure. However, if all the elements of this circuit are integrated upon one chip of silicon, the number of connections drops to approximately 14. All circuit elements are interconnected inside the package by a process known as *vapor metallization*; instead of 40 hermetic seals there is 1, and the 300 processing operations are reduced to approximately 30. Figure 3-2 presents a size comparison of a discrete J-K flip-flop circuit and an integrated circuit of the same type.

Before the actual fabrication of the integrated circuit begins, the silicon crystal must be sliced into paper-thin wafers (Fig. 3-3A). The wafers must be lapped and polished on the side that is to be used for the active elements. Unless special processing is involved, the back side of the wafer is left in the lapped state.

Both sides of the wafer are lapped simultaneously with an abrasive (usually aluminum oxide) until all visible traces of the saw cuts are removed. One side of the wafer is then polished several times with slurries of abrasive grit. A grit of smaller size is used for each succeeding polishing step. Finally, the wafer is chemically etched to remove any irregularities in the surface resulting from the last polishing step.

The diffusion process begins when the highly polished silicon wafer is placed in an oven, Fig. 3-3B, containing impurity atoms which yield the desired electrical characteristics. The concentration of impurity atoms diffused into the wafer can be controlled by regulating the temperature of the oven and the time that the silicon wafer is allowed to remain in the oven. When the wafer has been uniformly doped, the fabrication of semiconductor devices may begin. Several hundred circuits are produced simultaneously on the wafer, as shown in Fig. 3-4.

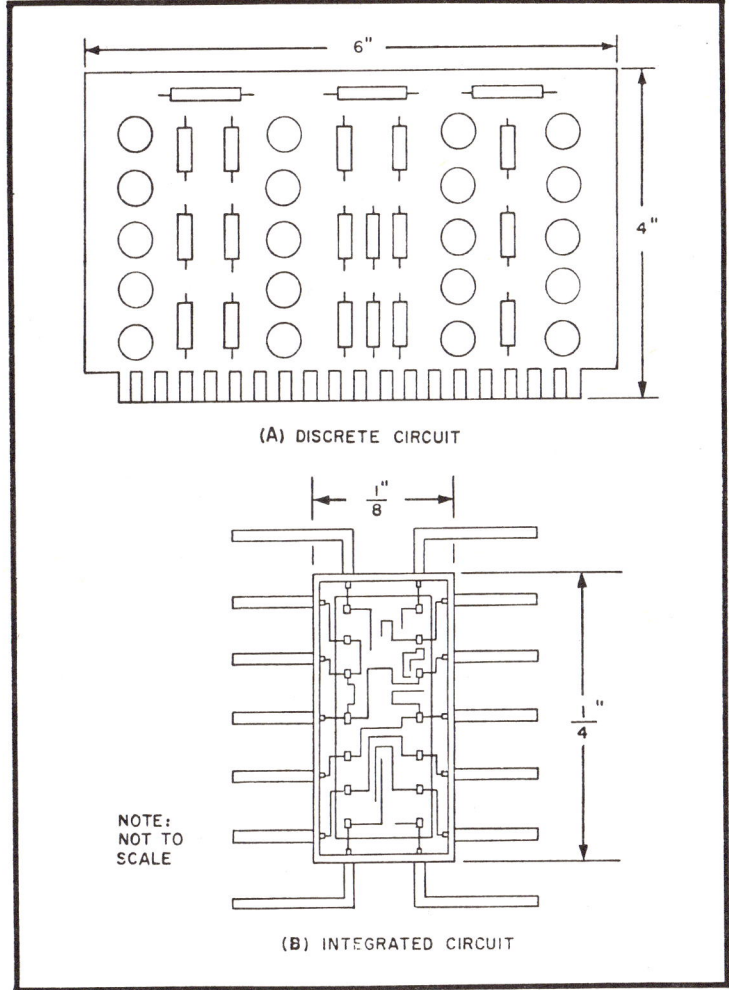

(A) DISCRETE CIRCUIT

NOTE:
NOT TO
SCALE

(B) INTEGRATED CIRCUIT

Fig. 3-2. Size comparison between a discrete and integrated circuit of a J-K flip-flop. The discrete circuit is approximately 5000 times larger.

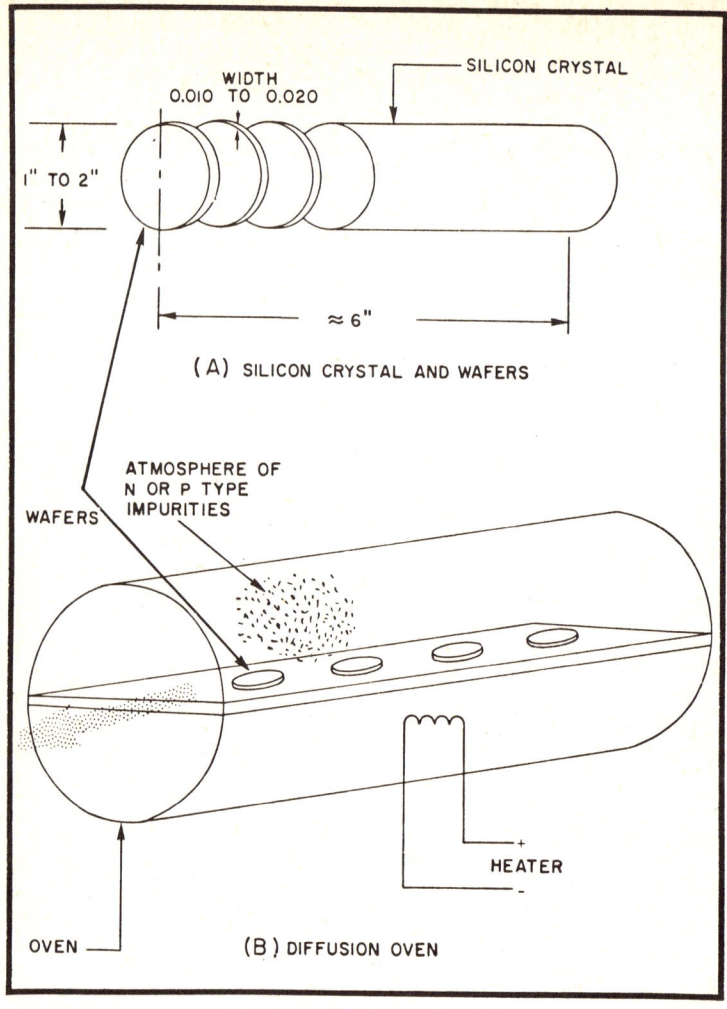

Fig. 3-3. Diffusion method.

THIN- AND THICK-FILM CIRCUITS

Film circuits differ from the integrated circuits in that instead of being diffused into the substrate, components are layered on the substrate material (usually ceramic) by deposition, screening, and etching, or a combination of these processes. Only passive devices are produced by this technique, since it does not lend itself to the production of active components. Active components, however, can be attached and interconnected in much the same way as leads are bonded to integrated circuits and transistors.

HYBRID CIRCUITS

A *hybrid* microcircuit is one that is fabricated by combining two or more circuit types, such as film circuits and semiconductor circuits, or even a combination of one or more circuit types with discrete elements (Fig. 3-5). The primary advantage of hybrid microcircuits over other microcircuits is design flexibility. (To avoid ambiguity in texts, device connections *inside* a given circuit package are called *intraconnections*; ordinary hookups *external* to the package are called *interconnections*.)

Several elements and circuits are available for hybrid applications. These include discrete components that are electrically and mechanically compatible with integrated circuits. They may be used to perform functions supplementary to ICs and can be handled, tested, and assembled with essentially the same technology and tools.

Complete circuits are available in the form of uncased chips (unencapsulated IC dice). These chips are usually identical to those sold as part of the manufacturer's regular production line. They must be properly packaged and connected by the user if a high-quality final assembly is to be obtained.

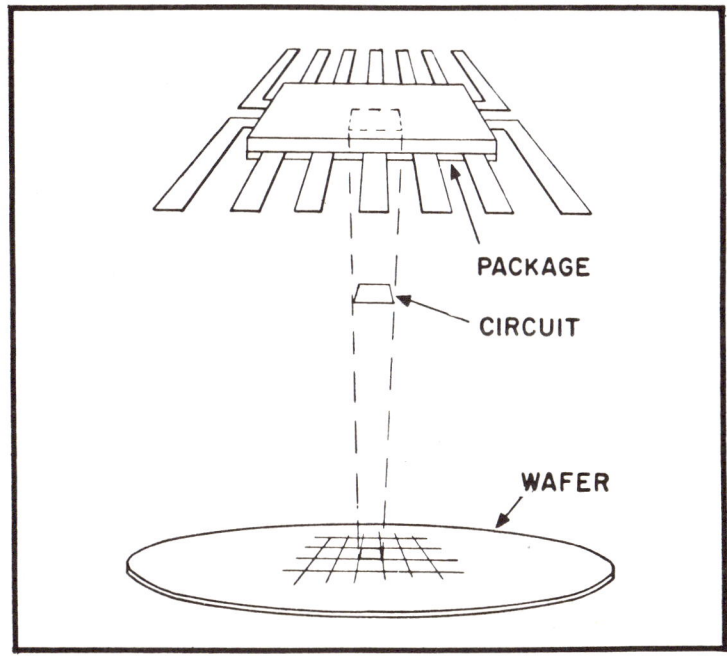

PACKAGE

CIRCUIT

WAFER

Fig. 3-4. Relationship of integrated circuit to wafer.

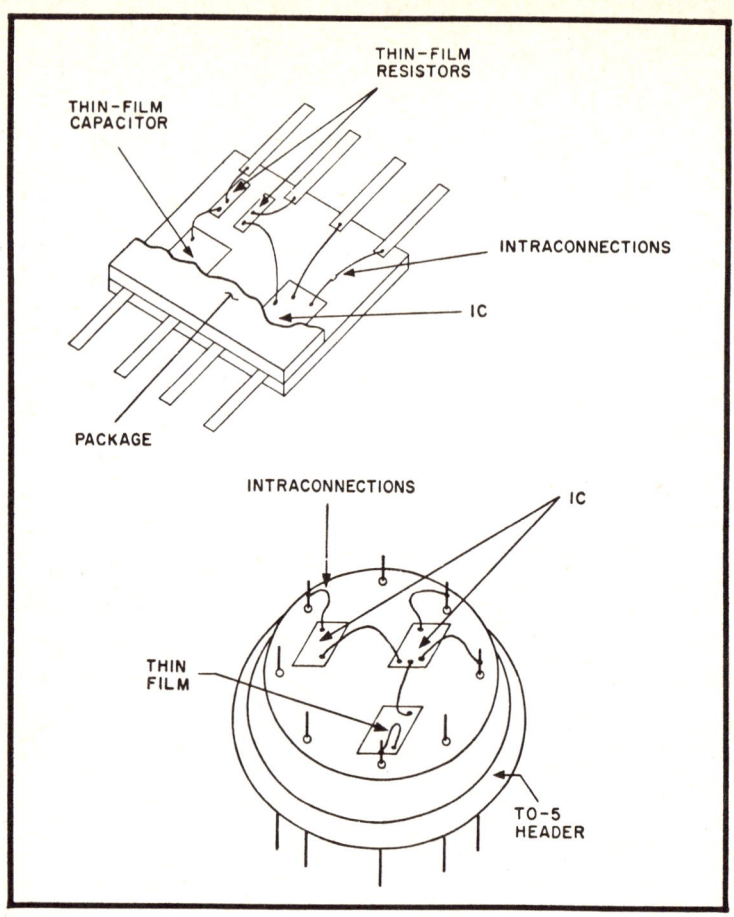

Fig. 3-5. Hybrid circuit configuration.

The circuits are usually sealed in a package to protect them from mechanical and environmental stresses. One-mil (0.001 in.) gold wire leads are connected to the appropriate pins which extend out of the package to allow external connections.

MICROELECTRONIC DEVICE PACKAGING

Integrated circuit packages have evolved from the still-used transistor type (modified to include more leads) to the widely used flat pack, and to the dual in-line package (DIP). Generally, these packages are limited to 14 or 16 leads, but some configurations are available with many more. Packages with up to 160 leads are being developed for large-scale integrated devices.

Modified TO-5 Package

Modified TO-5 is a term used to describe a package having the same general appearance as a standard TO-5 transistor package but which has been modified by increasing the number of external leads and the dimensions of the package.

Modified TO-5 assembly methods are extensions of the techniques used in the production of standard transistor packages. Since the modified TO-5 package typically has 8, 10, or 12 leads, compared with the 3 or 4 leads of the standard transistor package, its pin-circle diameter is slightly larger (0.230 in. vs 0.200 in.). The modified TO-5 package is shown in Fig. 3-6. Kovar, a nickel—iron alloy, is used for the leads and eyelet, and 7052 glass for the preform. The cover may be German silver, Kovar, or nickel; but Kovar is used most frequently because its coefficient of thermal expansion is similar to that of the glass preform, thus allowing a matched seal.

The glass preform and eyelet are assembled as shown in Fig. 3-6, and then sealed by fusion in a 1000°C oven. After it is sealed, the entire assembly, called the *header*, is cleaned, and the leads are clipped to the desired length. The header is then plated with 0.001 in. of gold cyanide. The IC die is bonded to the header by (1) inserting a solder preform between the die and the header, (2) placing a weight on top of the assembly, and (3) heating the entire header to approximately 390°C so that the die, solder, and header are fused together.

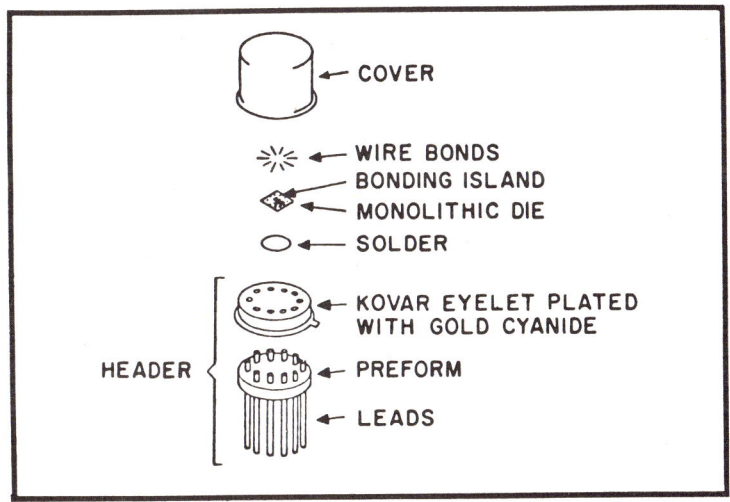

Fig. 3-6. Exploded view of modified TO-5 package.

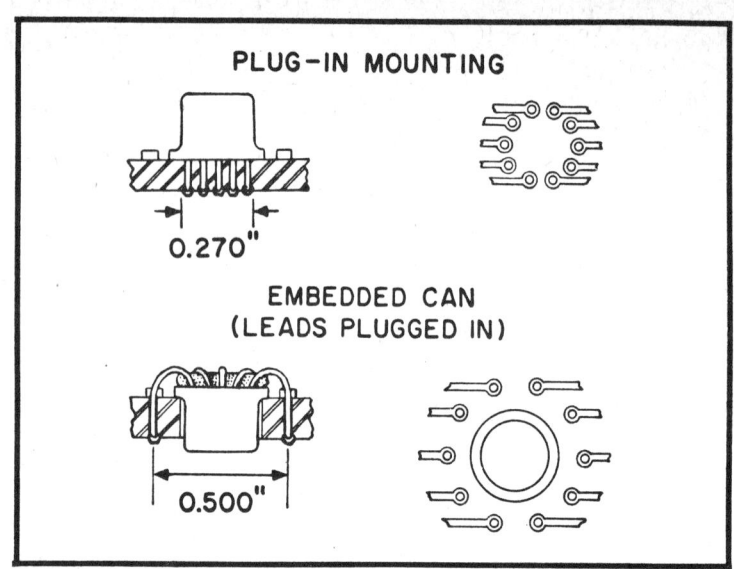

Fig. 3-7. Mounting of TO-5 packages.

Because the elements of integrated circuits (transistors, diodes, etc.) are so small, it would be impossible to attach wires directly to points in the circuit. For this reason, a series of relatively large bonding islands or pads are placed around the edges of the die during its fabrication; these are connected by metallization to appropriate points in the circuit. During packaging, the bonding islands are connected to the package leads by fine lead wires (0.001 to 0.003 in.). These wires may be either gold or aluminum, and are attached by thermocompression or ultrasonic bonding techniques. As the final step in TO-5 packaging, the cover is welded to the header in a controlled atmosphere.

The modified TO-5 package can be either pin-plugged or embedded into the printed circuit board (Fig. 3-7). The interconnection area of the package leads must not be impinged by other printed conductors on either side of the board. The plug-in method does not provide sufficient clearance between the pads to route additional circuitry between the leads, but when the packages are embedded, there is sufficient space between the pads (because of the increased diameter of the interconnection pattern) for additional conductors. The modified TO-5 package can also be plugged into a socket which has been soldered into the printed circuit board. Some sockets provide larger pin-circle diameters than the TO-5 package.

Flat Pack

Many types of integrated-circuit flat packages are being produced in various sizes and materials. These packages are available as square, rectangular, oval, and circular configurations with 10 to 60 external leads, and may be fabricated of metal, ceramic, epoxy, glass, or combinations thereof. Only the ceramic flat pack (Fig. 3-8) will be discussed here, since it is representative of all flat packs with respect to general package requirements.

After the package leads are sealed to the mounting base, a rectangular area on the inside bottom of the base is treated with metal slurry to provide a surface suitable for bonding the monolithic die to the base. The lead frame is then cut away from the secured leads. These leads and the metallized area in the bottom of the package are gold plated. The die is then attached by using a gold−silicon eutectic which is easily melted.

The die bonding step (Fig. 3-8) is followed by bonding gold or aluminum wires between the bonding island on the integrated circuit die and the inner portions of the package leads. Following the wire bonding, a glass−solder preformed frame is placed on top of the mounting base. One surface of the ceramic cover is coated with Pyroceram glass, and the cover is placed

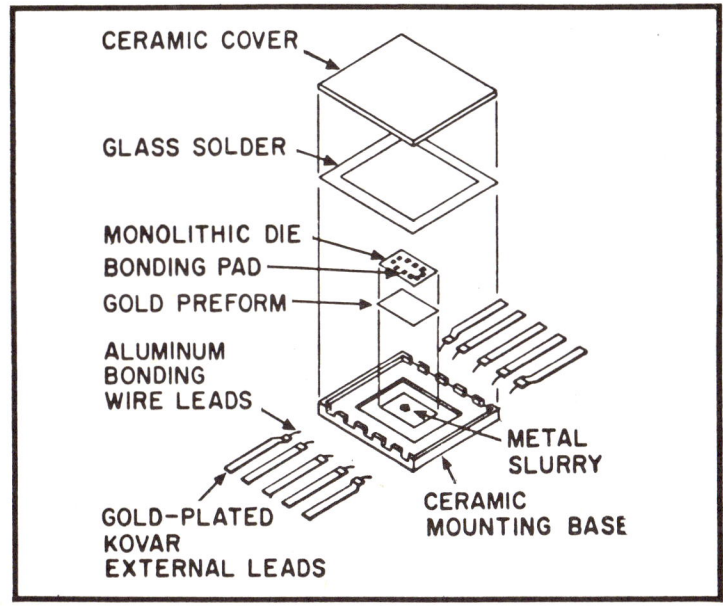

Fig. 3-8. Exploded view of ceramic flat pack.

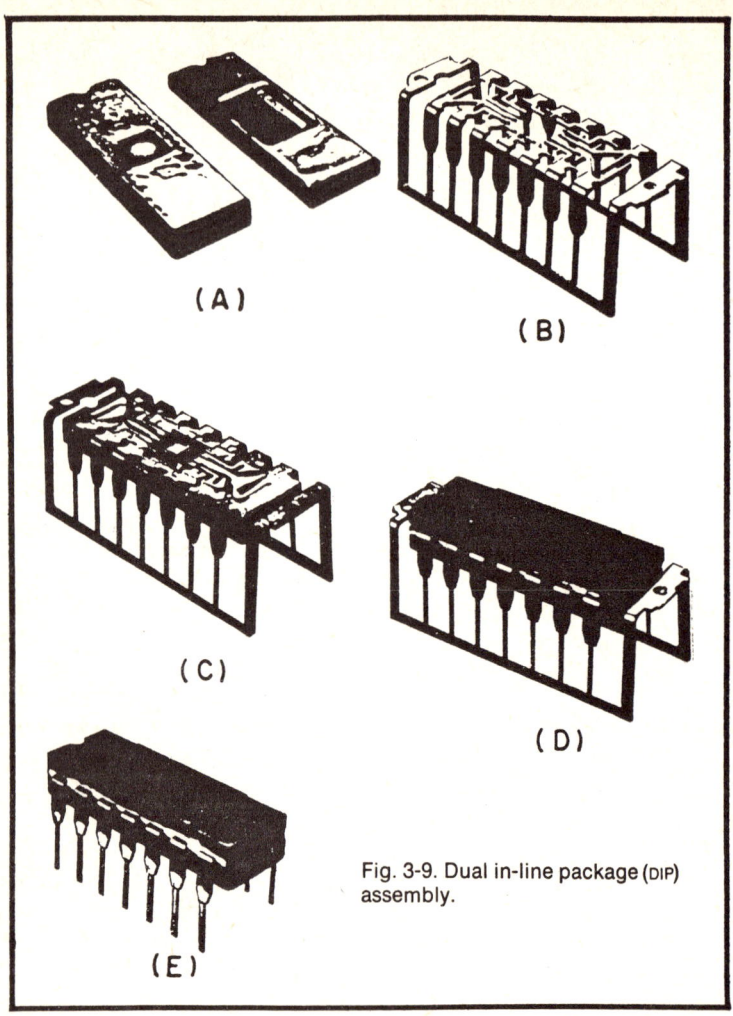

(A)

(B)

(C)

(D)

(E)

Fig. 3-9. Dual in-line package (DIP) assembly.

on top of the mounting base. The entire assembly is placed in an oven at 450°C. This causes the glass solder and Pyroceram to fuse and seal the cover to the mounting base.

Dual In-Line Package

The dual in-line package (DIP) was designed primarily to overcome the difficulties associated with handling and inserting packages into mounting boards. Dual in-line packages are easily inserted by hand or machine and require no spreaders, spacers, insulators, or lead forming. It is also possible to field-service the devices with standard hand tools and soldering irons. Plastic dual in-line packages are finding wide use in commercial applications, and a number of military systems incorporate ceramic in-line packages.

The progressive stages in the assembly of a ceramic dual in-line package are illustrated in Fig. 3-9. The integrated-circuit die is sandwiched between the two ceramic elements (A). The element on the left is the bottom half of the sandwich and will hold the integrated-circuit die. The ceramic section on the right is the top of the sandwich; it will protect the integrated-circuit die from mechanical stress during sealing operations. Each of the ceramic elements is coated with low-melting-temperature glass for subsequent joining and sealing.

The Kovar lead frame is shown stamped and bent into its final shape (B). The excess material is intended to preserve pin alignment. The holes at each end are for the keying jig used in the final sealing operation. The lower half of the ceramic package is inserted into the lead frame (C). The die is mounted in the well and leads are attached. The top ceramic element is bonded to the bottom element (D) and the excess material is removed from the package (E).

Ceramic dual in-line packages are processed individually while plastic dual in-line packages are processed in quantities of two or more, in chain fashion. After processing, the packages are sawed apart. The plastic package also uses a Kovar lead frame, but the leads are not bent until the package is completed. Because molded plastic is used to encapsulate the integrated circuit die, there may no void between cover and die as is the case with ceramic packaging. At present ceramic dual in-line packages are the most common of the two package types to be found in microelectronic systems.

The trend in microelectronic packaging is toward still higher circuit density. Complex arrays are already available to a limited degree, and research and development are being performed on extremely complex devices that will incorporate the required interconnections on silicon wafers. This approach, called *large-scale integration (LSI)*, has as its objective the interconnection of hundreds of undiced circuits packaged in containers comparable in size to a silver dollar.

Considerable effort is being directed toward increasing the upper frequency limit associated with microelectronics. The practicality of microwave integrated circuits has been demonstrated, and this type of circuit is currently being fabricated by a number of laboratories.

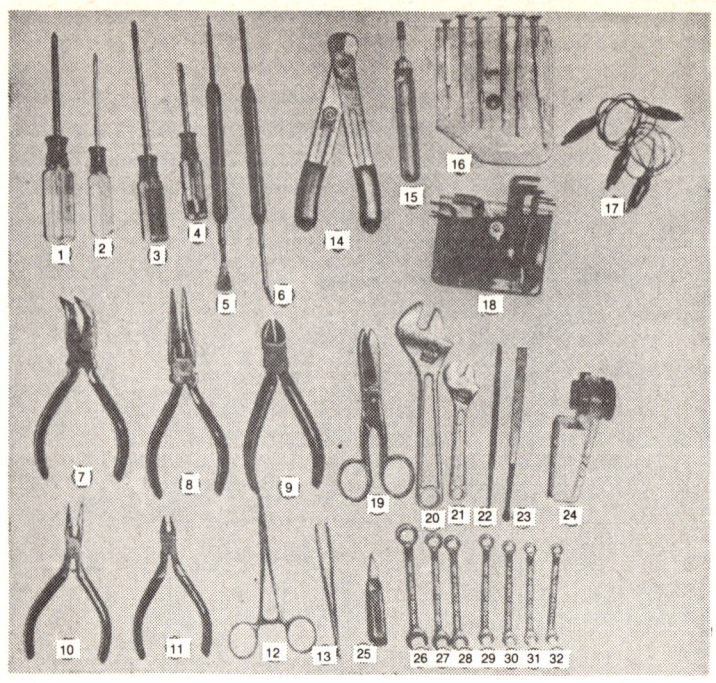

TOOL	DESCRIPTION	TOOL	DESCRIPTION
1.	PHILLIPS SCREWDRIVER, MEDIUM POINT	14.	INSULATION STRIPPER
2.	PHILLIPS SCREWDRIVER, SMALL POINT	15.	PRECISION OILER
3.	STRAIGHT SCREWDRIVER, 3/16" x 4"	16.	JEWELER'S SCREWDRIVERS
4.	STRAIGHT SCREWDRIVER, 1/8" x 2"	17.	JUMPER LEADS WITH MINIATURE ALLIGATOR CLIPS
5.	SOLDERING AID, WIRE BRUSH AND POINT	18.	ALLEN AND BRISTOL (SPLINE) WRENCH SET
6.	SOLDERING AID, ANGLED POINT AND SLOTTED TIP	19.	ELECTRICAL SCISSORS
7.	LONG NOSE PLIERS, ANGLED	20.	CRESCENT WRENCH, 6"
8.	LONG NOSE PLIERS, WITH WIRE CUTTER	21.	CRESCENT WRENCH, 4"
9.	DIAGONAL CUTTING PLIERS	22.	TRIANGULAR FILE
10.	NEEDLE NOSE PLIERS WITH WIRE CUTTER, MINIATURE	23.	IGNITION FILE
11.	DIAGONAL CUTTING PLIERS, MINIATURE	24.	12 X MAGNIFIER
12.	HEMOSTATS	25.	CLIP-ON HEAT SINK
13.	TWEEZERS	26-32.	COMBINATION (OPEN-END/BOX-END) WRENCH SET

Fig. 3-10. Tools for microelectronics repair.

GENERAL MAINTENANCE PRINCIPLES

Microelectronic developments have had a great impact on the test equipment, tools, and facilities that are necessary to maintain systems incorporating these developments.

Test Equipment

Early electronic systems could be completely checked out with general-purpose test equipment, but the time required to individually test the components in one of today's very complex systems would probably exceed the design life of the system. One improvement in system testing was accomplished by designing systems so that testing could be performed at various functional levels. This allows groups of components to be tested as a whole, thus reducing the time required to test components individually. One other advantage of this method is that complete test plans can be written to provide optimum sequencing of tests and signal waveshapes for each function level.

This method of testing also leads to the development of special support equipment, or test sets capable of simulating operating conditions of the system under test. Appropriate voltages or signals are applied by the equipment to the various functional levels of the system, and the output of each level is monitored. Testing sequences are prewritten and steps are switched in manually. The limits for each function level are preprogramed to give a go/no-go indication.

Another significant breakthrough for improved systems testing was the design of modular subassemblies equipped with input and output terminals that allow quick isolation of faults and easy location of defective parts. These modules are designed for easy removal from the system.

Although functional level testing and modular design have been successfully applied to most systems in use today, the trend toward increasing the number of subassemblies within a module by incorporating microelectronics will make this method of testing less and less effective. This has led to the present trend of developing both on-line and off-line automatic test systems. The on-line systems are being designed to continuously monitor performance and to automatically isolate faults to removable assemblies. Off-line systems automatically check out removable assemblies and isolate faults to the component part level.

When no automatic means of accomplishing fault isolation is available, general-purpose test equipment and good troubleshooting procedures may be employed, but such fault diagnosis should be attempted only by experienced technicians. Misuse of electrical probes or soldering irons may permanently damage printed circuit (PC) boards or the microelectronic devices attached to them. The proximity of leads to one another, and the effects of the interconnecting wiring makes the testing of PC boards extremely difficult, and drift or current leakage measurements practically impossible.

Printed circuit boards that have been *conformally* coated are difficult to probe because the protective coating is often too thick to penetrate for a good electrical contact and must be removed for electrical probe testing. Test leads with sharp needle points are available; these points easily penetrate the conformal coating to permit testing. Many PC boards, however, are designed with special test points that can be monitored either with special test sets or general-purpose test equipment. Another method of obtaining access to a greater number of test points is to use extender cards or cables.

If none of these test aids is available, each suspect device or component lead must be probed manually. However, special care must be exercised when probing integrated circuits, since the circuits may be easily damaged by excessive voltages or currents, and leads may be physically damaged.

Precautions concerning the use of test equipment for troubleshooting equipment containing integrated circuits are similar to those that must be observed when troubleshooting equipment containing semiconductor devices. To prevent possible circuit damage when testing microelectronic circuits with a VTVM or VOM, the meter should have a sensitivity of more than 20.000 ohms per volt on the voltage scales, and on the resistance scales should not pass more than one milliampere of current through the circuit being tested.

Static DC and resistive tests are usually effective in locating catastrophic failures or defects that exhibit large deviations from normal characteristics, but these methods are time-consuming and sometimes inadequate when the defect is a drift in the device characteristics. The suspect device in this case must be desoldered, then retested to verify the fault. If the defect is not verified, the device must be resoldered to the board. If it is necessary to repeat this procedure several times or if the printed circuit board is conformally coated, the defect may never be located. In fact, the circuit may be further damaged by the attempt to locate the fault. For these reasons the device should never be desoldered until all possible in-circuit tests are performed and the defect verified.

Tools

Because of the smallness and susceptibility to damage of microelectronic devices, the repair of modules incorporating them may be accomplished in many instances only with special tools designed expressly for that purpose. The following paragraphs describe some of these special tools which are also included in the recommended tool list (Table 3-1). The table is divided into three sections: Part I lists the special tools, and Part II the chemicals that are required to perform maintenance on systems incorporating microelectronic devices. Part III lists tools that generally are available to electronics technicians.

Table 3-1. Recommended Tools for Microelectronics Repair.

Item	Quantity	FSN or Mfg Part No	Manufactured by or Equivalent
Part I Special Tools			
Ungar Universal Handle	1	777	Ungar
Thread-in Element, 23½ Watts	1	535	Ungar
Thread-in Element, 37½ Watts	1	1235	Ungar
Thread-in Element, 47½ Watts	1	4045	Ungar
Pencil Tip	1	PL331	Ungar
Pencil Offset Tip	1	PL332	Ungar
Chisel Tip, Long Taper	1	PL333	Ungar
Tapered Needle Tip	1	PL338	Ungar
Stepped Pencil Tip	1	7154	Ungar
Ungar Hot Knife Tip	1	4025	Ungar
Offset Slotted Tip	1	862	Ungar
Straight Slotted Tip	1	857	Ungar
Hollow Cube Tip	1	863	Ungar
Cup Tip, ⅝" Dia.	1	856	Ungar
Cup Tip, ¾" Dia.	1	855	Ungar
Cup Tip, 1" Dia.	1	854	Ungar
Desoldering Tip (For DIPs)	1	859	Ungar
Desoldering and Cleaning Tool	1	7800	Ungar
Soldering Iron Holder	1	8000	Ungar
Kleen-Tip Sponge and Tray	1	400	Ungar
Low Voltage Soldering Iron		6970	Ungar
Printed Circuit Card Holder	1	371	Henry Mann Co.
Steel Bench Clamp	1	356	Henry Mann Co.
Miniature Vise	1	353	Henry Mann Co.
Illuminated Magnifier, 3 Power	1	LFM-1	Henry Mann Co.
Clip-on Lens, 4 Power	1	No. 1	Henry Mann Co.
Desoldering and Cleaning Tip, 0.033"	1	7812	Ungar
Hemostat	1	8-907	Fisher Scientific Co.
Hand tool—lead trimmer (TO-type package)	1		Henry Mann Co.
Desoldering and Cleaning Tip, 0.057"	1	7806	Ungar
Desoldering and Cleaning Tip, 0.069"	1	7813	Ungar
High Intensity Lamp	1	5975	Tensor
Lead Forming Tool	1	ATH-3260	Astro Tool Co.
Dual In-Line Package Puller	1	6982	Ungar
45° Chain-Nose-Tip Cutter	1	A119	Henry Mann Co.
Tooth Picks, Round	1 Box		
Wooden Dowels, ¼" × 6"	10		
Triceps Tweezer, 8"	1	T8	Henry Mann Co.
Part II Required Chemicals and Materials			
Insulating Varnish	1 Pt.	5970-280-4920	
Epoxy-Resin Compound	1 Kit	8040-777-0631	
Potting Syringes	12	E-602	Fishman
Epibond	1 Kit	H-1331	Furane Plastics
Acid Brushes	12		
Solvent Dispensers	4	613	Henry Mann Co.
Polyurethane Coating	1 Kit	CE-1155	Conap, Inc.
Isopropyl Alcohol	1 Pt.	TT-1-735	
Methyl-Ethyl Ketone	1 Pt.	TT-M-261	
Acetone	1 Qt.	6810-281-1864	
Copper Shim Stock	0.003"		
Aluminum Oxide Abrasive Paper	10 Shts/Box	5350-967-5080	
Cotton Swabs			Johnson & Johnson
Teflon Tape, 1"	1 Roll		
Soldering Aid Tool	1	G5120-629-2697	
End Nippers, 4½"	1	G5110-221-1503	
Diagonal Pliers, 4"	1	G5120-541-4079	
Long-Nose Pliers, 4½"	1	G5120-541-4078	
Round-Nose Pliers, 4½"	1	G5120-239-8252	
Phillips Screwdriver, 3"	1	G5120-240-8716	
Phillips Screwdriver, 4"	1	G5120-234-8913	
Phillips Screwdriver, 6"	1	G5120-234-8912	
Phillips Screwdriver, 10"	1	X1010	XCelite/Henry Mann Co.
Flat Tip Screwdriver, 2"	1	G5120-227-7377	
Flat Tip Screwdriver, 4"	1	G5120-278-1283	
Flat Tip Screwdriver, 5"	1	G5120-278-1271	
Flat Tip Screwdriver, 10"	1	R31610	XCelite/Henry Mann. Co.
Jeweler's Screwdrivers	1 Set	G5120-288-8739	
Swiss File Set	1 Set	G5110-288-7685	

Table 3-1. (Continued)

Part III Standard Tools			
Allen Wrenches	1 Set	G5120-288-8732	
Allen Wrenches	1 Set	G5120-315-3358	
Spin Tite, ¼"	1	G5120-770-0016	
Spin Tite, ⁵/₁₆"	1	5120-770-0017	
Spin Tite, ⅜"	1	5120-770-0025	
Spin Tite, ⁷/₁₆"	1	5120-770-0030	
Tweezers, Curved	1	G5120-288-9685	
Tweezers, Straight	1	G5120-247-0867	
Tweezers, Lock	1	G5120-293-0149	
Drill Bit Set 33 to 80	1	20-087	Hunter Tools
Hand Drill	1	F-223	Hunter Tools
Brass Wire Brush	1		
Heat Sinks, Medium	6	51F	Hunter Tools
Nylon Brush	1		
X-acto Knife (with Blades)	1	5110-595-8400	
Material Loot	1 Box		Johnson & Johnson
Solder Eutectic, 63/37 Rosin Core	1 Lb.		
Tool Box	1	7817	Union Steel Chest Corp.
Primary Page Teflon Tape, 1"	1 Roll		
Forceps, Straight, 5"	1	G6515-334-5600	
Forceps, Curved, 5"	1	G6515-334-4900	
Forceps, Straight, 9"	1	L6515-334-9900	
Forceps, Curved, 8"	1	L6515-334-4100	
"C" Clamp, 3 inch	1	5-810	Fisher Scientific Co.
Wire Stripper	1	101-S	Henry Mann Co.

Figure 3-10 shows some of the tools that are normally kept in the electronics technician's toolbox.

The most important tool used in removing and replacing microelectronic packages is the soldering iron. Improper choice of a soldering iron or tip may result in severe damage to microelectronic devices or printed circuit boards. Only a low-wattage pencil soldering iron and thin diameter solder with rosin core flux and a tin-to-lead ratio of 63:37 should be used.

Repair work on modules containing microelectronic devices usually involves both desoldering and resoldering. For desoldering, a soldering iron with interchangeable tips is recommended (Fig. 3-11). One tip designed specifically for heating and removing solder from leads has a hollow point through which molten solder is drawn. The vacuum for the point may be provided manually (using a suction ball) or mechanically (by means of a vacuum pump). The manually operated attachment is recommended because it requires no special equipment.

The *wicking (capillary) method* is an effective method for removing solder when no special equipment is available. In this method a piece of copper braid is applied to the melted solder, then removed. Due to capillary attraction, the copper braid acts as a wick absorbing the melted solder.

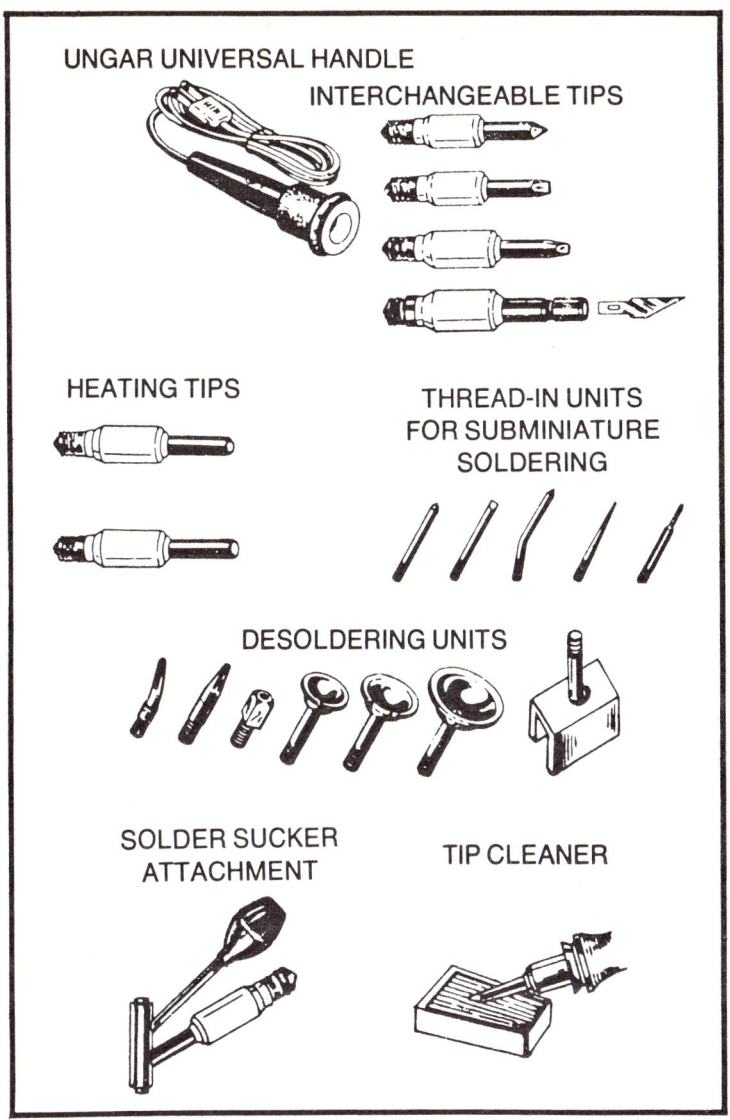

Fig. 3-11. Recommended soldering tools.

The other desoldering tips and heating elements shown in the figure are compatible with various types of microelectronic packages. The best way to remove a microelectronic package from a printed circuit board is to desolder all leads simultaneously. This can be accomplished only if a tip is available to accommodate the lead configuration of the package. Many tips are required because of the lack of standardization of microelectronic packaging.

Resoldering should be performed with a miniature isolated low voltage soldering system. Preferably, the heat of the tip should be voltage-controlled to provide the correct amount of heat to form a good solder joint. The small mass of the tip will assure that only the lead being soldered is affected by the heat from the iron. All soldering irons used in repair of modules containing microelectronic devices should be powered by isolated power supplies or be equipped with ground straps to eliminate stray voltages.

Another major item is a fixture for holding printed circuit boards while they are being repaired. This fixture should provide good support to the printed circuit board, but should also be easy to rotate since the technician, during the performance of maintenance actions, must often inspect or apply heat to both sides of the board.

In some cases the complete desoldering and resoldering operation and final inspection of a repair must be conducted under an illuminated magnifier. Soldering operations should be performed under a magnifier with a power of approximately ×3. Final inspection should be made with a ×12-power magnifier.

Special tools are also required for lead forming. Most integrated circuits are shipped in carriers or special lead frames to protect the devices against damage during handling or testing. Before the packages may be attached to a printed circuit board they must be removed from the carrier and the leads trimmed and formed. This may be accomplished in any of several ways. One way is to cut the leads of the integrated circuit and carrier to the desired length with an X-acto knife. Another method is to cut the leads with scissors or diagonal pliers. However, the physical shock of a lead being cut by a diagonal cutter may be transmitted along the lead, causing damage to the interior of the microelectronic device and causing an early failure. With either of these methods, some stress is placed on the integrated circuit leads that may eventually cause the device to fail.

An integrated-circuit forming tool is recommended for use in removing integrated circuits from their carriers. This tool has special dies that allow precise cutting and forming of leads in one operation with a minimum amount of stress on the integrated circuit leads.

After desoldering, most microelectronic devices can be removed from printed circuit boards with tweezers, but dual in-line packages require special extractor tools. The extractor tool listed in Table 3-1, Part I, is recommended to remove dual in-line packages and to prevent lead damage. Dual in-line packages are also difficult to insert if leads are bent.

Work Area

The work area for the maintenance and repair of microelectronic devices need not include special clean-room facilities or an exceptionally large work space. The main requirement is a clean, sturdy workbench, approximately the size of a standard desk and solid enough to accommodate such special equipment as a printed-circuit card holder, magnifier, small vise, etc.

A riser shelf should be provided for storage of general test instruments to be used for simple continuity tests. Tool holders should also be provided for larger tools. The more delicate instruments, such as tweezers, should be kept in a special tray or drawer.

Proper lighting is also important because of the extremely small size of integrated-circuit packages and the close tolerance requirements for lead alignment. Shadows and glare cause eye fatigue.

REMOVAL AND REPLACEMENT OF CONFORMAL COATINGS

Many manufacturers apply protective conformal coatings to their equipment. These conformal coatings include epoxies, silicones, polyurethanes, varnishes, and lacquers. Most of them consist of a synthetic resin dissolved in a volatile solvent. When properly applied to a clean surface, the solvent evaporates, leaving a continuous layer of solid resin. After curing, the coating protects against environmental stress, corrosion, moisture, and fungus. However, these coatings must be removed before maintenance can be performed on the equipment.

Conformal coatings can be removed either by chemical or mechanical means. However, the application of chemical solvents may damage PC boards by dissolving the adhesive materials that bond the circuits to the boards. The application of solvents may also dissolve the potting compounds used on other parts or assemblies.

Most polyurethane conformal coatings can be removed by applying heat to the area to be cleaned and then gently scraping the conformal coating with an X-acto knife. Detailed procedures

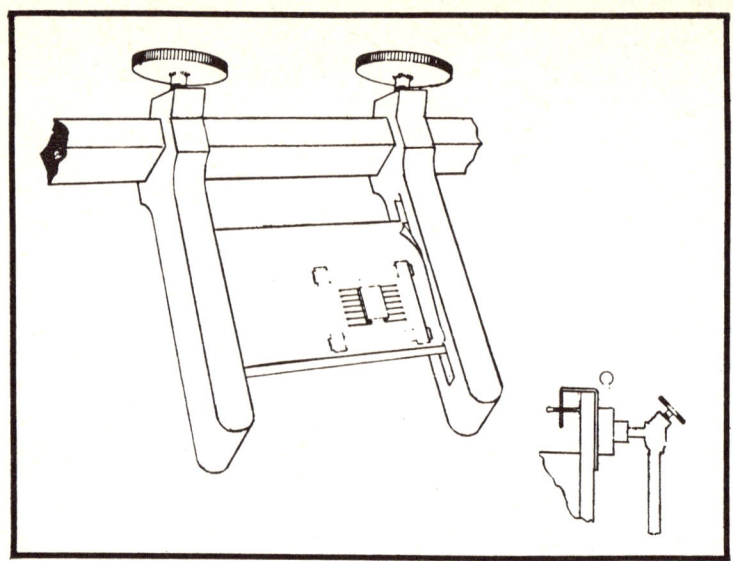

Fig. 3-12. Printed-circuit card holder.

for removing coatings by this method are presented in the following steps.

1—Place the PC board to be repaired in the PC board card holder as shown in Fig. 3-12.

2—Mask the area that is not to be stripped of conformal coating.

3—Avoid excessive heat, which may damage the PC boards or surrounding parts.

4—Heat a small area of the coating to be removed by holding a soldering iron close to the surface, but do not touch the board with the iron.

5—Do not apply excessive pressure during scraping, as this may cause nicks on the land of the PC board or damage to the leads of other parts.

6—When the conformal coating softens, gently scrape it away from the surface with an X-acto knife.

7—Remove loosened particles of conformal coating by gently brushing. See that loose particles of coating do not contaminate any moving parts.

8—Repeat these steps until all coating has been removed from the area exposed by the mask.

9—Cleanse the area with a cotton swab that has been dipped in alcohol to remove all loose particles of conformal coating that remain after brushing.

10—Remove the mask.

11—Cleanse the area with a cotton swab and alcohol—insuring that no loose particles of coating remain on the perimeter of the stripped area.

12—With a ×12 viewer (magnifier), visually examine the cleansed area to insure that the area is clean and that no damage was done during the cleaning operation.

To replace the coating after repair, proceed as follows:

1—With a ×12 viewer, visually inspect the repaired board for foreign particles.

2—Clean the area to be coated with a cotton swab that has been dipped in alcohol.

3—Allow the PC board to dry thoroughly. Drying time will depend on the cleaning agent used (alcohol is recommended).

4—Mask the area that is not to be coated. The unmasked area should be large enough to allow some overlapping with the old coating to assure adequate protection.

5—Spray the exposed area with three layers of conformal coating material, allowing sufficient time for curing between successive coats. Several thin coats are more effective than one heavy coat. If the coating material is slightly porous, it is unlikely that any pinholes will occur in exactly the same positions on successive coats.

6—Remove the mask and inspect the reworked area for any voids or pinholes, using a ×12 viewer. If any voids or pinholes are observed, add another coat of conformal coating.

Conformal coatings may affect the capacitance of air-dielectric capacitors, the distributed capacitance between leads, and the Q of inductors in RF assemblies. Refer to the appropriate maintenance manual for alignment procedures and the proper use of conformal coatings in RF assemblies.

REPAIR OF PRINTED CIRCUIT BOARDS

Printed circuit wiring patterns are formed in three basic ways: by painting, chemical deposit, and stamped or etched metal foil. Repair of painted and chemically deposited printed wiring patterns is not recommended, because the specialized equipment needed is not generally available.

The metal-foil printed wiring pattern consists of a thin metal foil bonded to a nonconductive base. The wiring pattern is

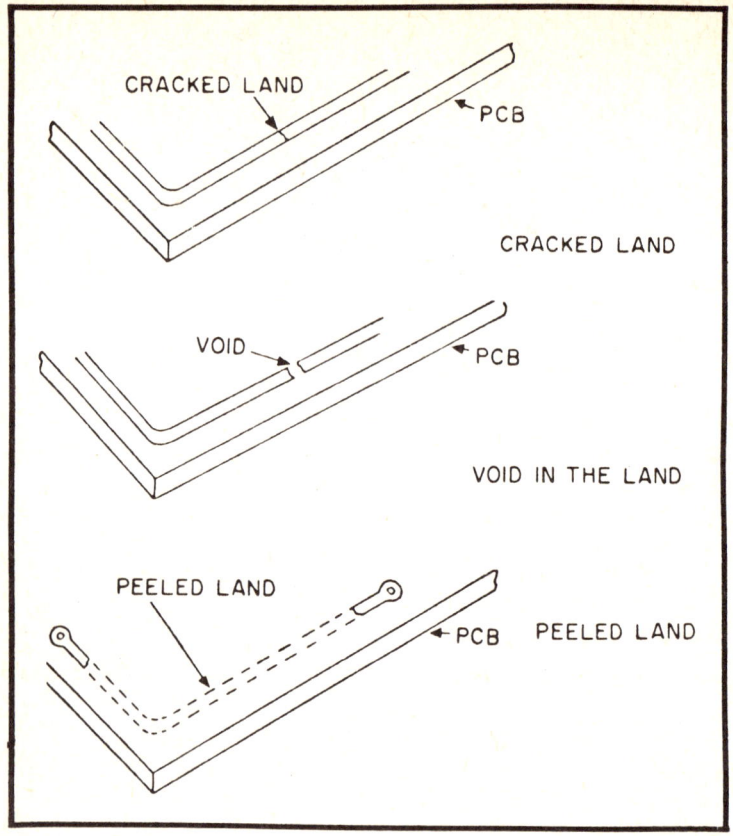

Fig. 3-13. Examples of broken conductor lands.

produced by stamping the foil before bonding it to the base, or by chemically etching away unwanted portions of the metal foil after bonding to the base. Because the metal foil is the most readily repairable and most commonly used type of printed wiring board, the repair techniques described in this section apply to metal-foil printed circuits only.

The major cause of PC board failures, attributable to maintenance actions, is mishandling when isolating faults and replacing parts. It is important, therefore, that care be exercised in performing maintenance.

After isolation of a fault to a PC board, the board should be visually examined to determine the possible cause of the fault. If the cause is not readily apparent, the lands on the board should be checked for continuity with an ohmmeter and needle probes. Place the probes at each end of the land. If the land is open, move one probe along the board until continuity is observed on the meter. Then locate the break in the land with the ×12 viewer.

The three major types of PC board failures are cracks, voids, and peeling of the lands, as shown in Fig. 3-13. A cracked land on a PC board may be located under magnification. These types of damage can be repaired by following the procedures described in the following paragraphs.

The following tools and materials are required:

>Miniature low-voltage soldering iron
>General-purpose adhesive such as Hysol
>Epoxy or Epibond H-1331
>Alcohol, isopropyl
>Acetone
>Cotton swabs
>Teflon tape
>C-clamp
>Flat file (Swiss file)
>Copper shim material
>X-acto knife
>Aluminum oxide sandpaper
>Ohmmeter
>Teflon-coated hookup wire, AWG 22
>×3 viewer
>Potting syringe

Repair of Cracked Lands

1—If the land has been covered with a conformal coating, remove it in accordance with the procedure described earlier.

2—Flow-solder the cracked sections together as shown in Fig. 3-14.

3—Clean all residue from the boards with a cotton swab that has been dipped in alcohol.

4—With a ×3 viewer, visually inspect the area for solder flux and solder splashes. Remove any residue.

5—Replace conformal coating as described earlier.

6—Check continuity.

Insure that solder does not flow over the edges of the land as this will reduce the spacing between lands and affect the electrical performance characteristics of the board.

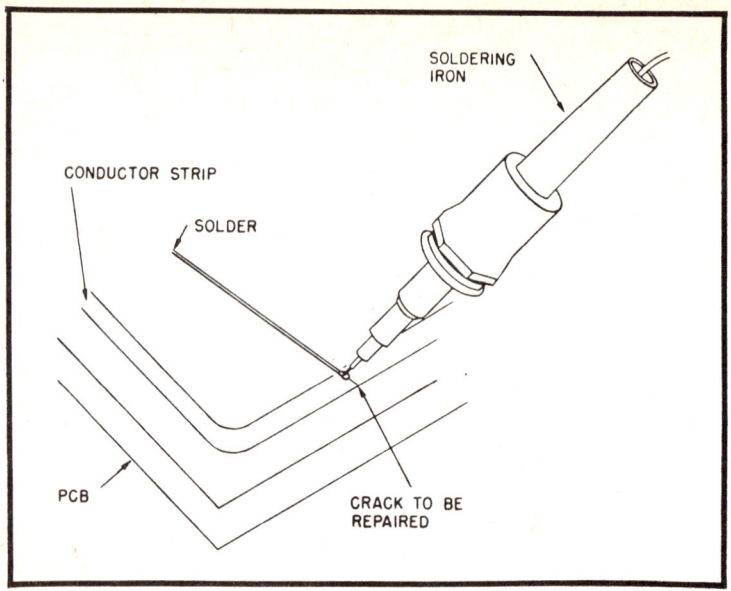

Fig. 3-14. Repair of cracked lands.

Repair of Small Voids

1—Remove the conformal coating as stated previously.

2—With an X-acto knife, trim each end of the broken land at a 45° angle (Fig. 3-15A). Insure that the remaining end of each land is firmly attached to the board.

3—Strip the insulation from a small piece of AWG 22 hookup wire and solder as shown in Fig. 3-15B.

4—Clean the soldered areas with cotton swabs that have been dipped in alcohol.

5—With a ×3 viewer, visually inspect the area for solder flux and solder splashes. Remove any residue.

6—Check continuity.

7—Replace conformal coating as described previously.

Excessive heat may cause the ends of the conductor to lift from the board.

Repair of Peeled Conductors

1—Remove the conformal coating as described previously.

2—With an X-acto knife, trim each end of the land at a 45° angle. Insure that the remaining end of each land is firmly attached to the board.

3—Obtain a length of AWG 22 hookup wire with Teflon insulation that will span the void in the conductor. Strip both ends and tin.

4—Position the wire as shown in Fig. 3-16 and solder the ends to the terminals.

5—Pot the wire to the board with general-purpose adhesive to prevent breakage during handling of the board.

6—With a ×3 viewer, visually inspect the area for solder flux and solder splashes. Remove any residue.

7—Check continuity.

8—Replace conformal coating as described previously.

Repair of Loose Connector Tabs

1—Place the edge of an X-acto knife between the board and the loose tab, lifting the tab as shown in Fig. 3-17.

2—Place a piece of Teflon on the top of the tab and clean the underside of the tab with aluminum oxide sandpaper until it is a bright color.

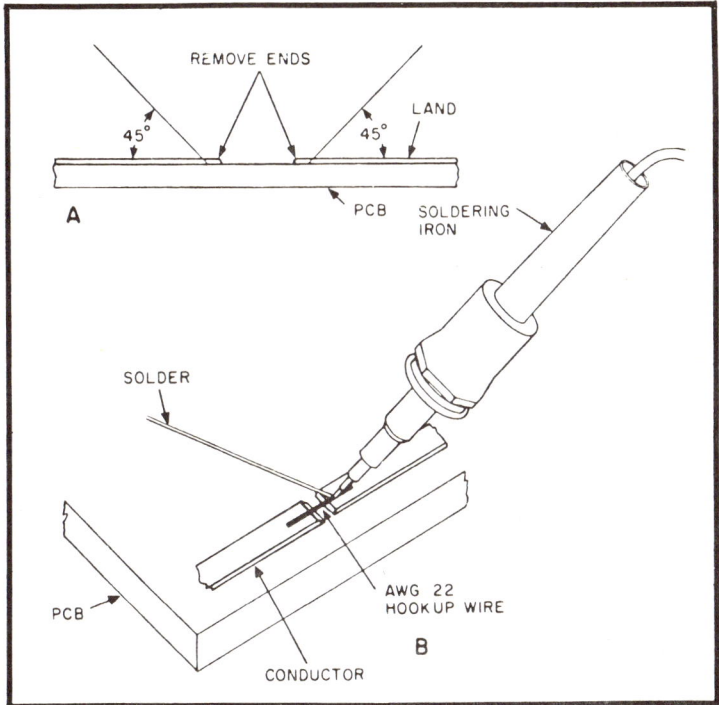

Fig. 3-15. Repair of land with void.

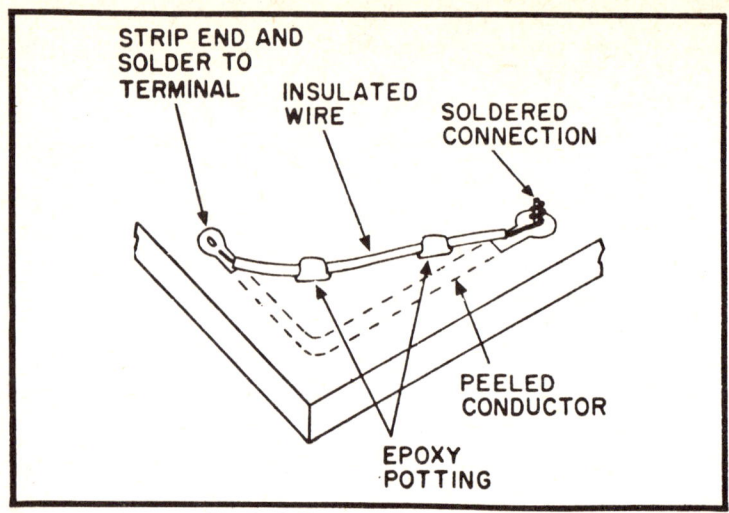

Fig. 3-16. Repair of lifted land.

3—Remove all old adhesive from the board by gently scraping with an X-acto knife.

4—Gently wipe the underside of the tab and board with cotton swabs that have been dipped in acetone.

5—Using a potting syringe, apply adhesive to tab and board surfaces.

Do not allow the tab to curl or form right angle bends as this will damage the tab. Acetone is flammable.

6—Place Teflon on top of the tab and clamp both to the board with C-clamp.

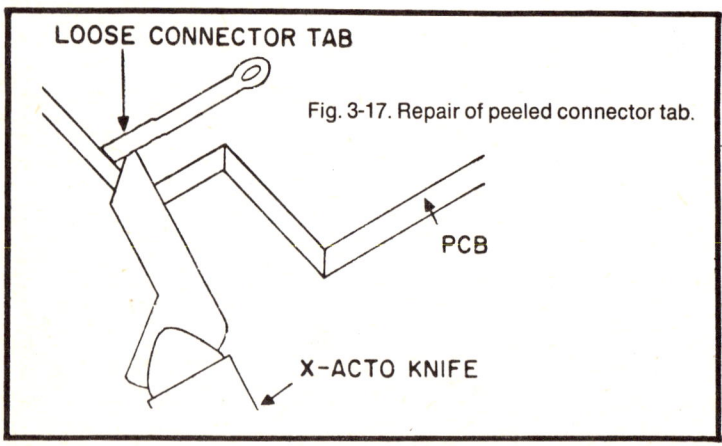

Fig. 3-17. Repair of peeled connector tab.

7—Allow adhesive to cure in accordance with the manufacturer's specifications.

8—Remove C-clamp and Teflon.

9—Gently scrape the board with an X-acto knife to remove excess adhesive.

10—Clean tab by wiping with a cotton swab that has been dipped in acetone.

11—Insure that the repaired contact fits into the connector.

12—Check continuity.

Repair of Broken Connector Tabs

1—With an X-acto knife, cut off the rough edge of the broken tab by making a 45° angle cut as shown in Fig. 3-18.

2—Insure that the undamaged section of the tab is still bonded to the board.

3—Clean the board by scraping off any residue and wiping with a cotton swab that has been dipped in acetone.

4—Cut a replacement contact from a piece of 0.002 in. thick copper foil stock. Cut to the same dimensions as the tab to be replaced, allowing ¼ in. for a lapjoint at the point of contact with the existing copper land.

5—Clean the replacement contact by holding it with tweezers and dipping it in acetone. Place the cleaned contact on a clean piece of paper.

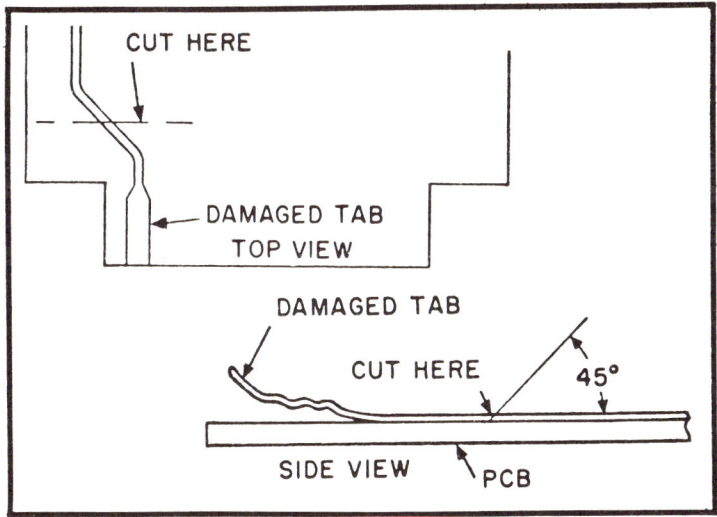

Fig. 3-18. Damaged connector tab.

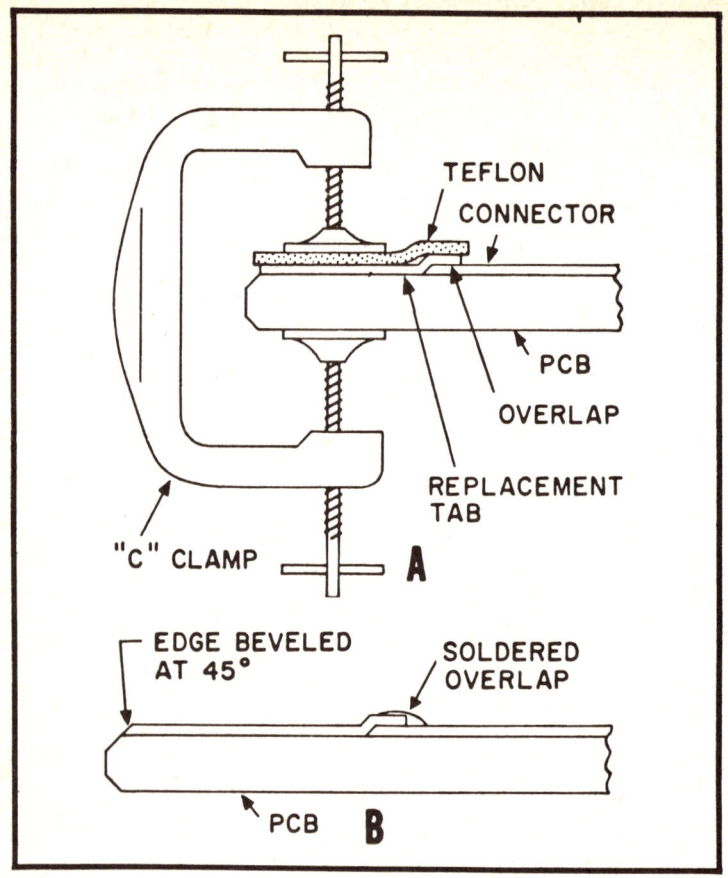

TEFLON
CONNECTOR

PCB

OVERLAP

REPLACEMENT
TAB

"C" CLAMP **A**

EDGE BEVELED
AT 45°

SOLDERED
OVERLAP

PCB **B**

Fig. 3-19. Connector tab replacement.

6—Use the potting syringe to prepare adhesive in accordance with the manufacturer's directions and apply it to the board.

7—Align the replacement contact with tweezers, and press flat.

8—Using an X-acto knife, scrape off excess adhesive, leaving a thin film on the board.

9—Place a piece of Teflon over the replaced tab and clamp with a C-clamp as shown in Fig. 3-19A. Do not move Teflon or rotate clamp while tightening, as the movement may result in improper orientation of the tab.

10—Allow adhesive to cure in accordance with the manufacturer's specification. To decrease curing time, place a lamp (60W to 100W) approximately 5 in. from the board.

11—Remove C-clamp and Teflon. Solder at the point of overlap between the new and original circuitry, using a miniature low-voltage soldering iron.

12—With a small flat file, bevel the edge of the tab at a 45° angle to conform to the bevel of the board (Fig. 3-19B).

13—Gently scrap the board with an X-acto knife to remove excess adhesive.

14—Clean contact and soldered joint by wiping with a cotton swab that has been dipped in alcohol.

15—Insure that the repaired contact mates evenly with the proper connector.

16—Check continuity.

REMOVAL AND REPLACEMENT OF FLAT PACKS

Because of their small size, extreme care must be exercised in soldering and unsoldering flat-pack leads. Careless work may cause damage to the mounting surfaces, circuits, or both.

Orientation of the flat pack with respect to the mounting surface is also of major importance. All flat packs have index points, usually located in one corner or on the package centerline. Before removing any flat pack, the board to which it is attached should be marked, or a sketch made of the location of the index point so that the replacement device may be properly oriented (Fig. 3-20).

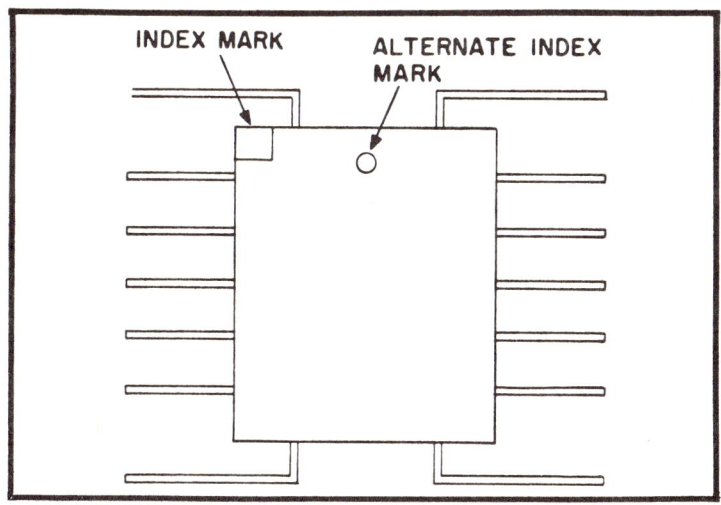

Fig. 3-20. Flat pack index markings.

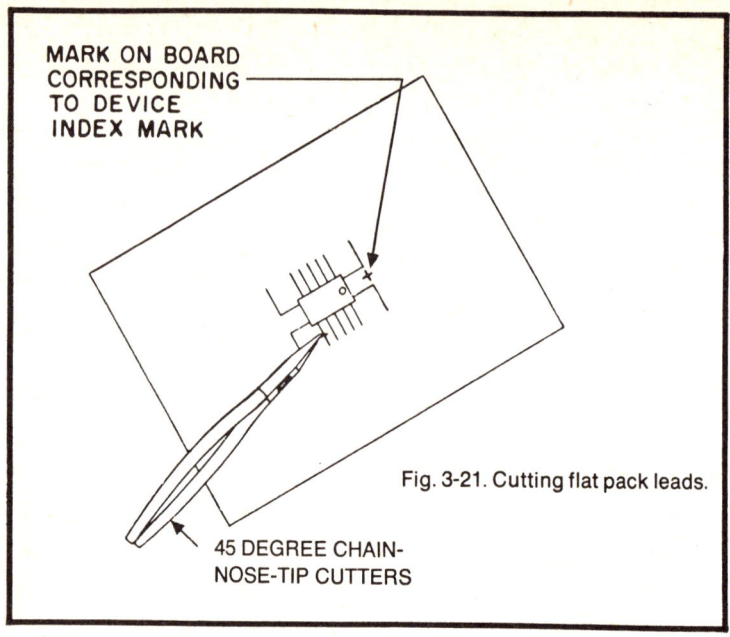

MARK ON BOARD
CORRESPONDING
TO DEVICE
INDEX MARK

Fig. 3-21. Cutting flat pack leads.

45 DEGREE CHAIN-
NOSE-TIP CUTTERS

The following tools and materials are required:

45° chain-nose-tip cutter
Desoldering and cleaning tool (solder sucker)
Low-voltage soldering iron
X-acto knife
Printed-circuit card holder
Teflon tape
Epoxy resin compound
×3 and ×4 viewers
Alcohol, isopropyl
Cotton swabs
Tricep tweezers
Lead cutting and forming tool (or wooden form)

Soldered-In Flat Packs

Upon completion of subsystem checkout and the localization of the defective flat pack, the following removal and replacement procedures are recommended.

1—Insert the PC board in PC card holder with the flat pack face up as shown in Fig. 3-21.

Solvents and abrasives may damage the board or surrounding parts and are not recommended for field maintenance. They should be used only under controlled conditions.

2—If the flat pack has been covered with a conformal coating, remove the coating as described for printed circuit boards.

3—Mark board to show the location of the index mark on the flat pack (Fig. 3-21).

4—With a pair of sharp chain-nose-tip cutters, cut the flat pack leads halfway between the soldered joints and the body of the flat pack as shown in the figure.

5—If thermally conductive adhesive has been used between the flat pack and the board, hold the device with tricep tweezers. Place the heated blade of an X-acto knife between the flat pack and the printed circuit board on one side of the flat pack and gently move the knife back and forth in a cutting motion. Repeat this process on all four sides until the flat pack has been loosened.

Do not tap board to remove excessive solder, as this may cause bridging of lands on other parts of the board.

6—Unsolder the remaining leads, one at a time, using a desoldering iron and solder sucker attachment to remove excess solder.

7—After removing the flat pack, remove mask and visually examine the PC board with a ×12 viewer.

8—Remove all residue (conformal coating, solder flux, and solder splashes) and clean the board with a cotton swab that has been dipped in alcohol.

Insure that cleaning solvent is dry before applying a hot soldering iron—alcohol is flammable.

9—Examine the replacement device for damage. Review applicable circuit specifications and test circuit in accordance with the applicable service manual.

10—Remove the flat pack from the handling container. If a suitable lead cutting and forming tool is not available, place the flat pack on a wooden form and cut the leads to the proper length, one at a time, with a sharp X-acto knife (Fig. 3-22).

Insure that the replacement flat pack circuit is properly oriented with respect to the index mark previously scribed on the printed circuit board.

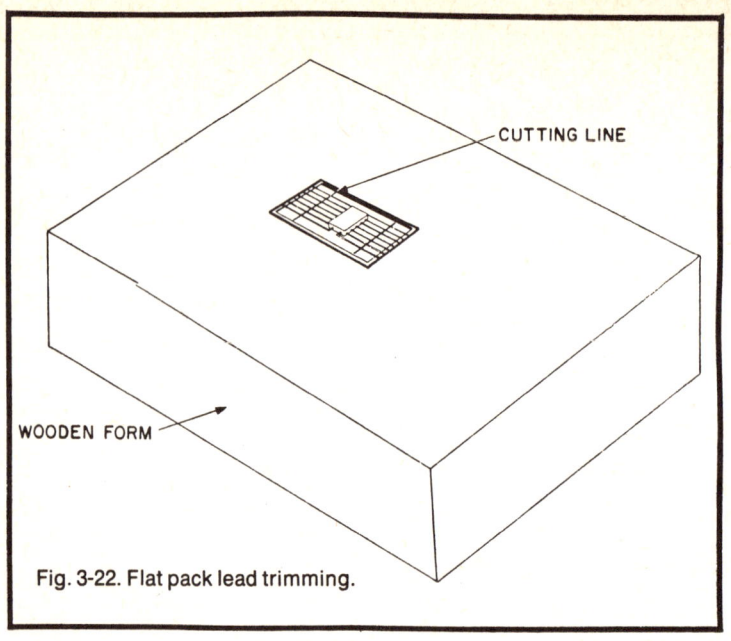

Fig. 3-22. Flat pack lead trimming.

11—Using the potting syringe, apply epoxy resin compound to the underside of the flat pack and place the package in position, using tricep tweezers.

12—Allow the adhesive to cure in accordance with the manufacturer's requirements.

Use a miniature low-voltage soldering iron. Excessive heat will damage the flat pack and the printed circuit board.

13—Flow-solder the flat pack leads to the board, one at a time. Use a miniature low-voltage iron (Fig.3-23).

14—With a ×12 viewer, visually examine the area surrounding the replaced device for bad solder joints, bridging of solder between the leads, flux residue, or excessive adhesive.

15—Clean all residue from the leads and PC board with a cotton swab that has been dipped in alcohol.

16—If the system is available, place the PC board in it for final test and checkout. If the system is not available, the proper inputs should be duplicated in accordance with specified test procedures and the output monitored using bench test equipment and procedures in the appropriate service manual.

17—Replace conformal coating as described previously.

Welded-In Flat Packs

Welded-in flat packs are removed and replaced by following the same procedures as for the soldered-in flat packs. Those portions of the leads remaining on the printed circuit board after removal of the package are clipped as closely as possible to the welded joint to facilitate soldering of the lapped joint of the replacement leads.

REMOVAL AND REPLACEMENT
OF DUAL IN-LINE PACKAGES

Because of their smallness, extreme care must be exercised in soldering and unsoldering dual in-line package (DIP) leads. Careless work may damage the mounting surface or the circuits.

Orientation of the DIP with respect to the mounting surface is also of major importance. All DIPs have index points, usually located in one corner or on the package centerline. Before removing any DIPs, the board to which it is attached should be marked or a sketch made of the location of the index point so that the replacement device may be properly oriented (Fig. 3-24).

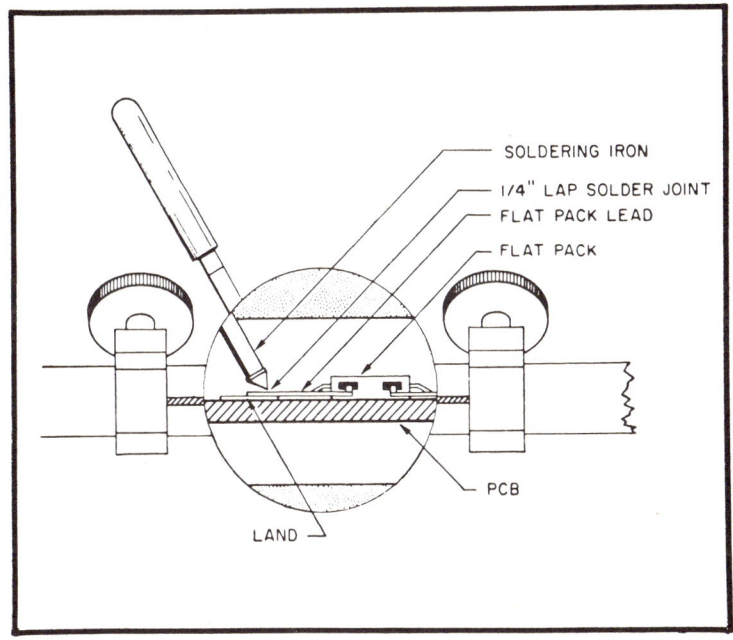

Fig. 3-23. Lap soldering of flat pack leads.

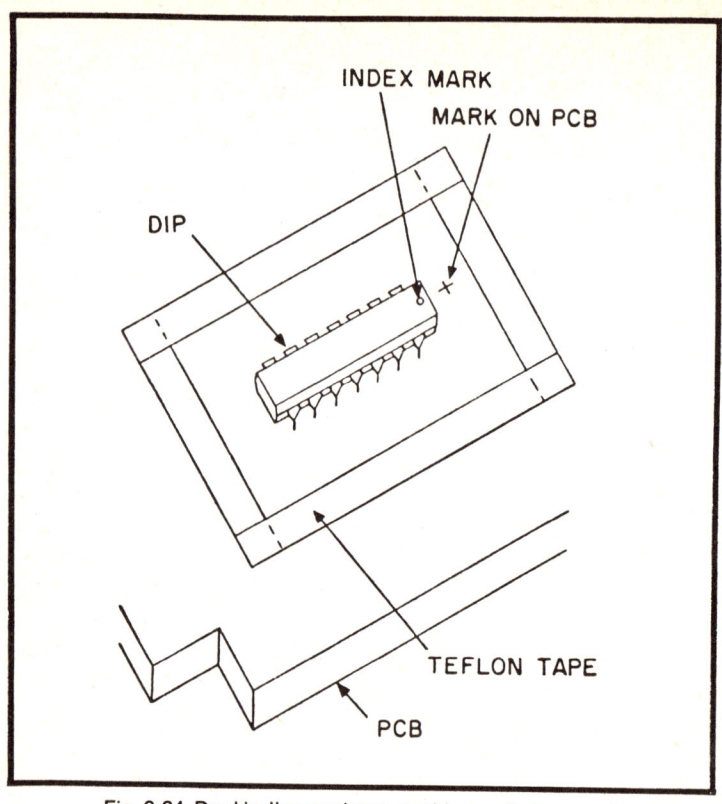

Fig. 3-24. Dual in-line package masking and alignment.

The following tools and materials are required:

45° chain-nose-tip cutter
Desoldering and cleaning tool (solder sucker)
Low-voltage soldering iron
Solder sucker attachment
X-acto knife
Printed-circuit card holder
Teflon tape
Epoxy resin compound
×3 and ×4 viewers
Alcohol, isopropyl
Cotton swabs
Tweezers, straight
Insulating varnish
Dual in-line package puller
Toothpicks

DIPs With Conformal Coating

1—Insert the board in a PC card holder, as shown in Fig. 3-12.

2—Construct a Teflon tape mask over the PC board, exposing the DIP to be removed. Remove the conformal coating.

3—Mark the board or make sketch indicating location of the index mark on the DIP.

4—With a pair of 45° chain-nose-tip cutters, cut the package leads, one at a time (Fig. 3-25).

5—Lift the portion of the leads attached to the package, one at a time, with tweezers, bending the leads upward.

Do not apply a prying or lifting motion with X-acto knife as this may damage the printed circuit boards.

6—Using a heated X-acto knife blade, loosen the conformal coating beneath the package and gently lift the package.

7—Remove the remaining conformal coating from the board.

8—Remove solder from each lead remaining in the board, using soldering iron and solder sucker and heater. Simultaneously remove the leads with tweezers.

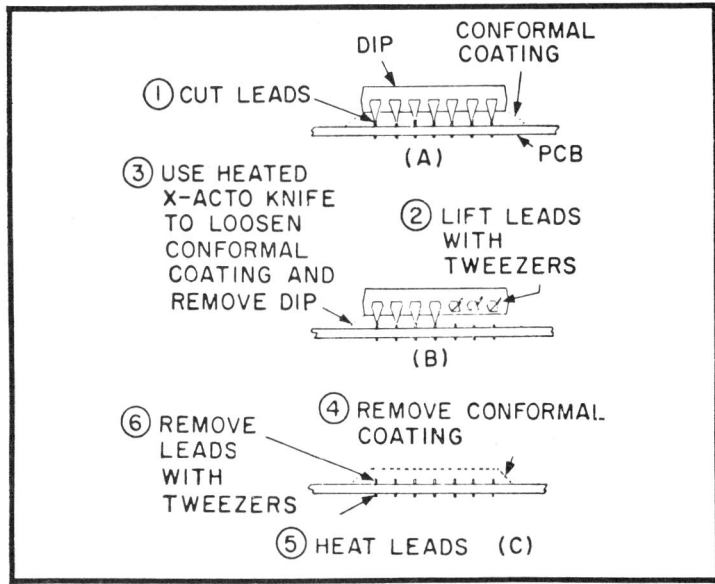

Fig. 3-25. Dual in-line package lead cutting and package removal.

9—Clean holes in the PC board with a toothpick.

10—Remove all residue (conformal coating, solder flux, and solder splashes) and clean board with a cotton swab that has been dipped in alcohol.

11—Examine the replacement device for damage. Review the applicable circuit specification and test circuit in accordance with the applicable technical manual.

12—Upon completion of the visual and electrical tests, align the circuit pins.

Insure that the circuit is properly oriented. Improper positioning of the package may result in destruction of the circuit.

13—Insert the DIP in the board and solder leads individually. Use a low-voltage soldering iron.

14—Use a ×12 viewer to visually inspect the area surrounding the replaced device for bad solder joints, bridging of solder between leads, and flux residue.

15—Clean all leads and surrounding areas with a cotton swab that has been dipped in alcohol.

16—Place the PC board in the system, if the system is available, for final test and checkout. If the system is not available, the proper inputs should be duplicated and the outputs monitored using bench test equipment and procedures in the appropriate manual.

DIPs Without Conformal Coating

1—Mark printed circuit board, showing location of the index mark on the DIP.

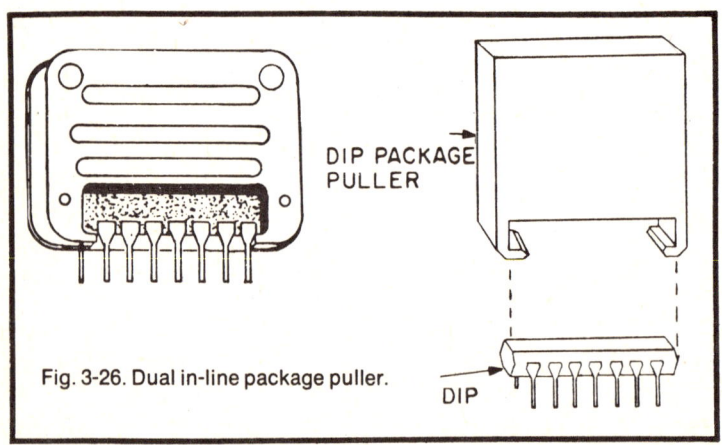

DIP PACKAGE PULLER

Fig. 3-26. Dual in-line package puller.

DIP

2—Heat solder at each lead, individually, and remove the molten solder from top of board with a solder sucker or wicking tool.

Do not apply twisting or prying forces, as this may damage the printed circuit board or break dual in-line package pins within the board

3—Grasp the DIP with the DIP puller illustrated in Fig. 3-26. Heat all leads simultaneously, using the DIP desoldering tip.

4—Gently pull the DIP away from and perpendicular to the board. The DIP should be easily freed from the printed circuit board.

5—Clean holes in the printed circuit board with a toothpick.

6—Remove all residue (solder flux and solder splashes) and clean board with a cotton swab that has been dipped in alcohol.

7—Examine the replacement device for damage. Review the applicable circuit specification and test circuit in accordance with the applicable technical manual.

8—Upon completion of the visual and electrical tests, align the circuit pins.

Insure that the circuit is properly oriented. Inproper positioning of the package may result in destruction of the circuit.

9—Insert the DIP in the board and solder leads individually, using low-voltage soldering iron.

10—With a ×12 viewer, visually inspect the area surrounding the replaced device for bad solder joints, bridging of solder between leads, and flux residue.

11—Clean all leads and surrounding area with a cotton swab that has been dipped in alcohol.

12—Place the PC board in the system, if the system is available, for final test and checkout. If the system is not available, the proper inputs should be duplicated and the outputs monitored using bench test equipment and procedures in the appropriate equipment test manual.

REMOVAL AND REPLACEMENT OF TO-TYPE PACKAGES

TO packages containing integrated circuits are available with 8 to 12 leads. The leads are usually arranged in a

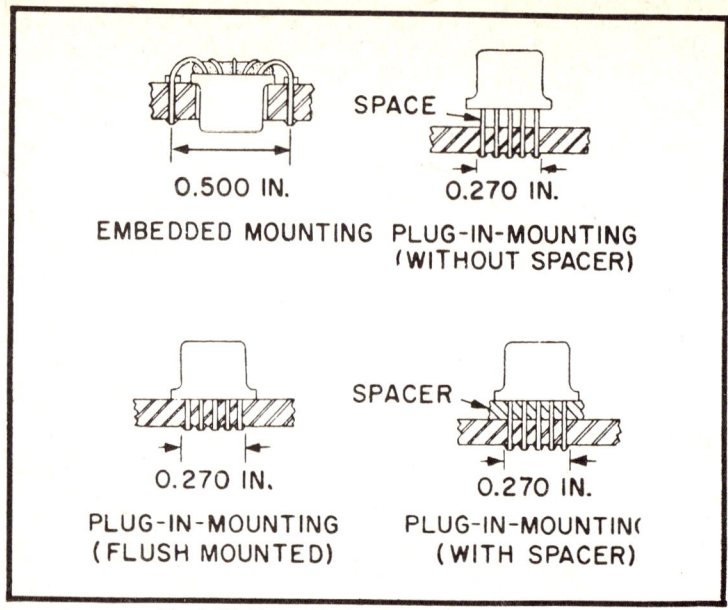

Fig. 3-27. Embedded and plug-in mountings.

symmetrical pattern and mounted directly to lands on the printed circuit board. Two of the most common mounting techniques, embedded and plug-in, are shown in Fig. 3-27.

The procedures currently used to remove and replace standard TO packages also apply to modified TO packages containing integrated circuits. The major difference between the two packages is that the number of leads on modified TOs is greater and the space between leads is less than on standard TOs, thus limiting the space available for lead clipping, desoldering, and soldering. The smaller packages also require greater manual dexterity on the part of repair personnel. Other constraints such as spacing of lands on the PC board, removal of conformal coatings, etc. present additional problems to the maintenance man.

Protective conformal coatings contribute to the difficulty in unsoldering leads, removing devices from the mounting surface, and preparing the mounting surface for device replacement. After the protective coating has been removed, device leads must be disconnected and the device removed, but the procedure to be followed in removing the TO will depend on the mounting configuration. If the package to be removed is embedded or plugged in without a spacer (Fig. 3-27), the leads should be clipped and the package removed before those segments of the leads remaining in the board are unsoldered. If this is not possible—as would be the case with packages that are flush-mounted or plugged in with a spacer—all leads should be heated simultaneously to allow package removal.

Procedures for removing embedded, plug-in, and flush-mounted packages are presented in the following paragraphs. Replacement TO circuits should be electrically tested in accordance with the applicable detailed specifications and subsystem operating requirements.

The following tools and materials are required:

> Teflon tape
> Desoldering and cleaning tool (solder sucker)
> Low-voltage soldering iron
> 45° chain-nose-tip cutter
> Long-nose pliers
> Printed-circuit card holder
> Toothpick
> Hemostat
> Alcohol, isopropyl
> Cup tip
> Cotton swabs
> ×3 and ×4 viewers

Embedded TOs and TOs Without Spacers

1—Place the PC board in the PC card holder.
2—If the PC board has been covered with a conformal coating, mask all portions of the board except for the TO that is to be removed.

Excessive heat may damage the TO package or printed circuit board. Solvents or abrasives should not be used. They may damage the printed circuit board or surrounding parts.

3—Remove coating from the terminal areas and area where the package is inserted into the board. This can best be accomplished by holding a soldering iron close to the printed circuit board and gently scraping the conformal coating with an X-acto knife.
4—Clip leads near the TO package with a 45° chain-nose-tip cutter.
5—Apply pressure to top of TO package with a wooden dowel to remove TO from the board (Fig. 3-28).
6—Heat terminals with a desoldering iron and remove molten solder with a solder sucker attachment.

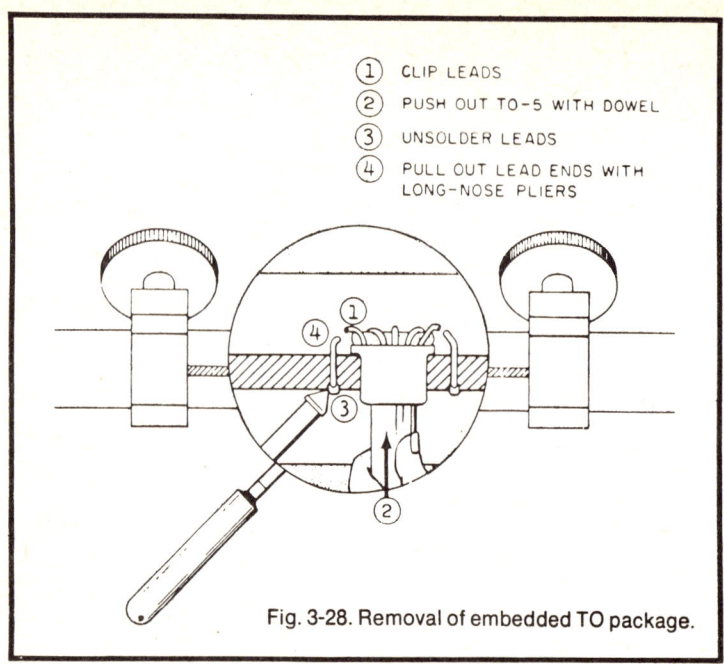

1. CLIP LEADS
2. PUSH OUT TO-5 WITH DOWEL
3. UNSOLDER LEADS
4. PULL OUT LEAD ENDS WITH LONG-NOSE PLIERS

Fig. 3-28. Removal of embedded TO package.

Do not forcibly pull or apply twisting motion to leads. This may damage terminals.

7—Remove those segments of leads remaining in the board by applying the desoldering iron to the printed circuit board land and gently pulling downward on the lead with long-nose pliers.

Wait until alcohol is dry before using soldering iron—alcohol is flammable.

8—Clean all terminal areas with a cotton swab that has been dipped in alcohol.

9—With a ×12 viewer, inspect terminals for loose solder, solder flux residue, and damage to terminals.

10—Visually examine replacement device for damage, review applicable circuit specifications, and test replacement circuit in accordance with requirements of the appropriate specification.

11—Align leads and insert new TO package into printed circuit board.

12—Solder leads individually with a miniature low-voltage soldering iron, using a heatsink or long-nose pliers as a heatsink.

13—Inspect repaired area with a ×12 viewer and remove any solder splashes or foreign materials with a cotton swab that has been dipped in alcohol.

14—If required by the applicable equipment maintenance manual, apply conformal coating.

15—Place the repaired PC board in the system, if the system is available, for final test and checkout. If the system is not available, the proper inputs should be duplicated and the outputs monitored using bench test equipment and following procedures in the appropriate equipment test and checkout manual.

Flush-Mounted TOs and Plugged-In TOs With Spacers

1—Place the printed circuit board in printed card circuit holder.

2—If the printed circuit board is covered with a conformal coating, mask all portions of the board except for the TO that is to be removed.

3—Heat each lead individually and remove molten solder from the printed circuit board with the solder sucker attachment.

Do not apply twisting motion to package. This may damage printed circuit board or terminals.

4—Heat all leads simultaneously with the cup-type tiplet adapter attached to the desoldering iron shown in Fig. 3-29. With hemostats, gently grasp the package and remove from the printed circuit board.

Wait until alcohol is dry before using soldering iron—alcohol is flammable.

5—Clean all terminal areas with a cotton swab that has been dipped in alcohol.

6—With a ×12 viewer, inspect terminals for loose solder, solder flux residue, and damage to terminals.

7—Visually examine replacement device for damage. Review applicable circuit specifications and test circuit in accordance with requirements of the appropriate operational specification.

8—Align leads and insert new TO package in printed circuit board.

9—Solder leads, one at a time, with a miniature low-voltage soldering iron.

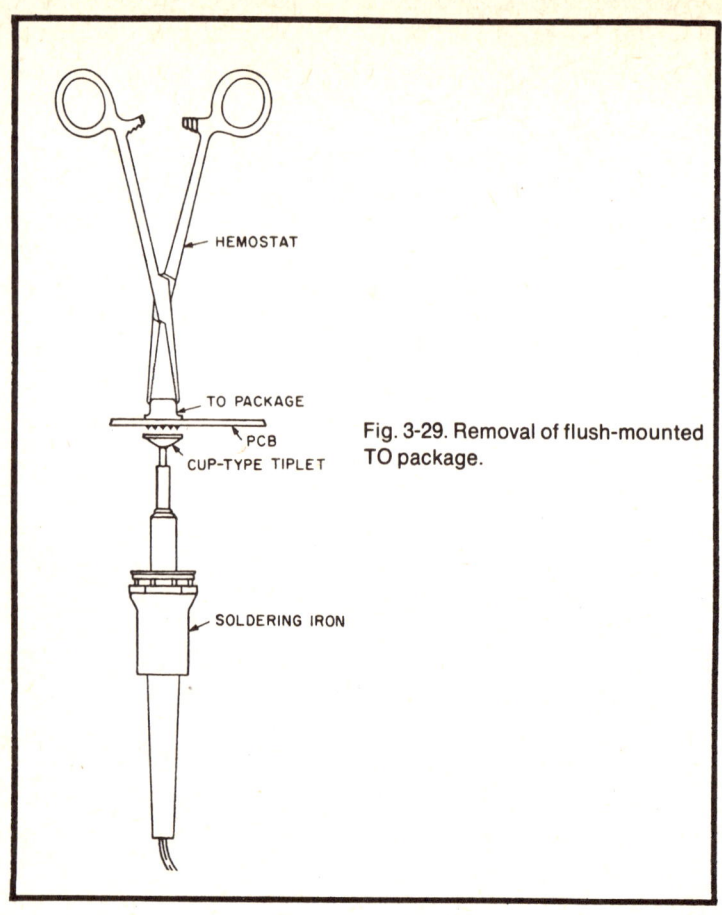

Fig. 3-29. Removal of flush-mounted TO package.

10—Inspect repaired area with a ×12 viewer and remove any solder splashes or foreign materials with a cotton swab that has been dipped in alcohol.

11—If required by the applicable equipment maintenance manual, apply conformal coating.

12—Place the PC board in the system, if the system is available, for final test and checkout. If the system is not available, the proper inputs should be duplicated and the outputs monitored using bench test equipment and following procedures in the appropriate equipment manual.

Chapter 4

Digital Logic Circuits and Binary Arithmetic

There are two main categories into which integrated circuits may be placed: *digital* and *linear*. Most integrated circuits are in the digital category, which includes the ICs used in computers, calculators, and other electronic "brains." In view of the preeminence of digital ICs, it is fitting that we first consider them in detail. At this point, all that needs to be said about the other category of ICs is that linear ICs are basically amplifiers.

Digital ICs are basically switching circuits used to make simple "yes or no" decisions. These decisions are made in accordance with certain laws, which form the basis of *digital logic*. Large numbers of digital ICs are combined in computers according to the laws of digital logic to perform the complex tasks required of computers.

Early computers used the same number system as you and I: the ordinary decimal system which, of course, has 10 digits. This was awkward because it meant that 10 different voltage levels had to be used to represent the set of decimal digits. Consequently the *binary number system* was adopted for computers and other digital systems.

The binary system has but two digits: 1 and 0. These are relatively easy to represent. In fact, an ordinary switch can represent a 1 when open and a 0 when closed. Since transistors can be turned on and off, they can be used as switches to represent the binary digits in digital logic circuits.

THE DIGITAL COMPUTER

The principle of operation of a digital computer is basically the process of counting. The ability to make calculations is built

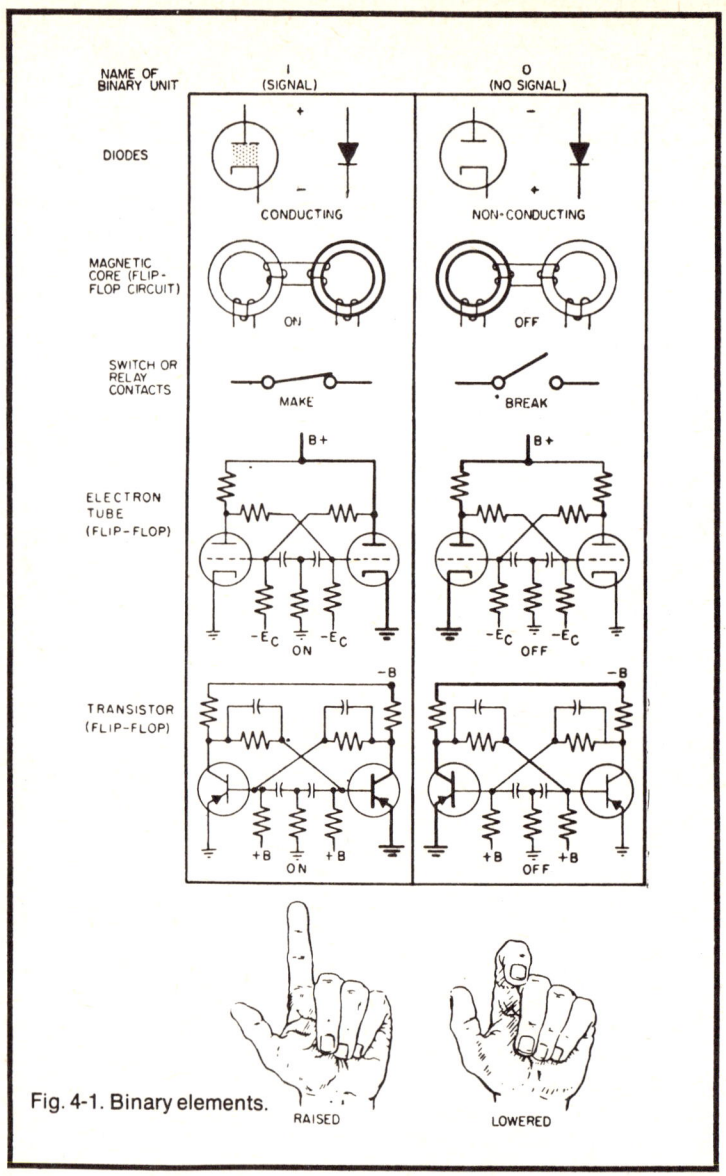

Fig. 4-1. Binary elements.

into the computer by using appropriate combinations of relatively simple electronic circuits. These circuits include components such as diode rectifiers, transistor flip-flops, and magnetic cores, all of which have two possible states (*bistable*). They are either on or off, conducting or nonconducting, energized or deenergized.

There are two types of digital computers: special purpose and general purpose. The special-purpose digital computer is designed to do a limited number of jobs and has a *fixed memory*.

The general-purpose, or *stored program*, digital computer is capable of many jobs. Instructions are stored in the memory unit by a process called *read-in*. These instructions may be changed by reading in a new set of instructions. Thus, the general-purpose digital computer has great flexibility.

NUMBER SYSTEMS

Digital computers function in the binary number system, using digits 0 and 1. The components that represent data can only represent two possible stable states, as previously stated. Figure 4-1 shows some of the binary elements used in digital computers.

Another basic unit which we personally can use as a counter is the finger. It can do only one of two things: it can be raised or lowered. No other finger positions can be considered to have any meaning. If it is raised, we say that it has value; if it is lowered, it has no value. It is therefore a binary element. We are fortunate in that we ordinarily have ten of these binary units, providing us with a ready-made binary digital computer.

By using only the fingers of our two hands, we can count to any number between 0 and 1023. We can do this by assigning values to the fingers of each hand. We call this number-assignment process *coding*. Let the thumb of the right hand represent 1; then the second finger represents twice the value of the first, or 2. The third finger represents twice the value of the second, or 4. Continue assigning each finger of the right hand twice the value of the one before it, as shown in Fig. 4-2. Then assign each finger of the left hand a number as indicated in the drawing. Note that each value we assigned to a finger is twice as great as the one preceding it. A quick look at the number sequence shows that the terms in the sequence are powers of the number 2. Thus

$$1 = 2^0 \qquad 8 = 2^3$$
$$2 = 2^1 \qquad 16 = 2^4, \text{ and so on.}$$
$$4 = 2^2$$

Weighting Values

It is more convenient to use symbols instead of fingers to represent binary numbers. A raised finger can represent a 1, and a 0 can be represented by a lowered finger. Remember, we

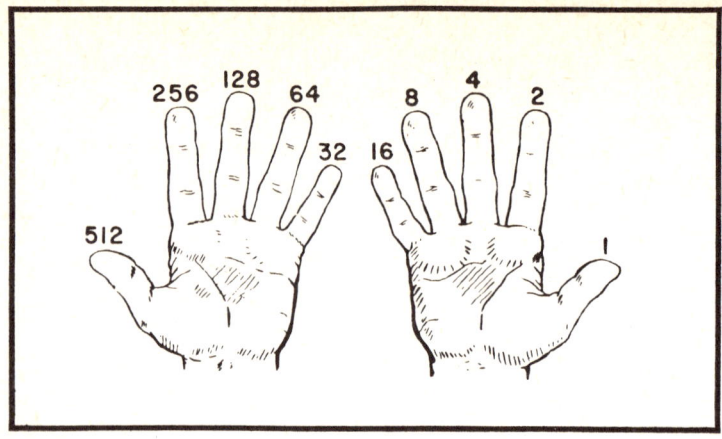

Fig. 4-2. The 8-4-2-1 code.

can do this because the finger is a binary unit; it can give two states. The value of each finger position is shown in the table in Fig. 4-3. This value is called the *digit position weighting value*, *weighting value*, or *weight* for short.

Consider the top binary number, 0000010101. We can find its value simply by adding the weighting value of each position that has 1 below it. Adding from right to left, we obtain:

$$21 = 1 + 4 + 16$$

POWERS OF 2	2^9	2^8	2^7	2^6	2^5	2^4	2^3	2^2	2^1	2^0		DECIMAL VALUE
WEIGHTING VALUE	512	256	128	64	32	16	8	4	2	1		
	0	0	0	0	0	1	0	1	0	1	=	21
	0	0	0	0	0	0	0	0	0	0	=	0
	1	1	1	1	1	1	1	1	1	1	=	1023
	1	0	0	0	1	1	1	0	0	0	=	568
	0	0	0	0	0	0	0	1	0	1	=	5
	0	0	1	0	0	0	0	1	0	1	=	133
	0	0	0	0	0	0	0	0	0	1	=	1

Fig. 4-3. The digit-position weighting value.

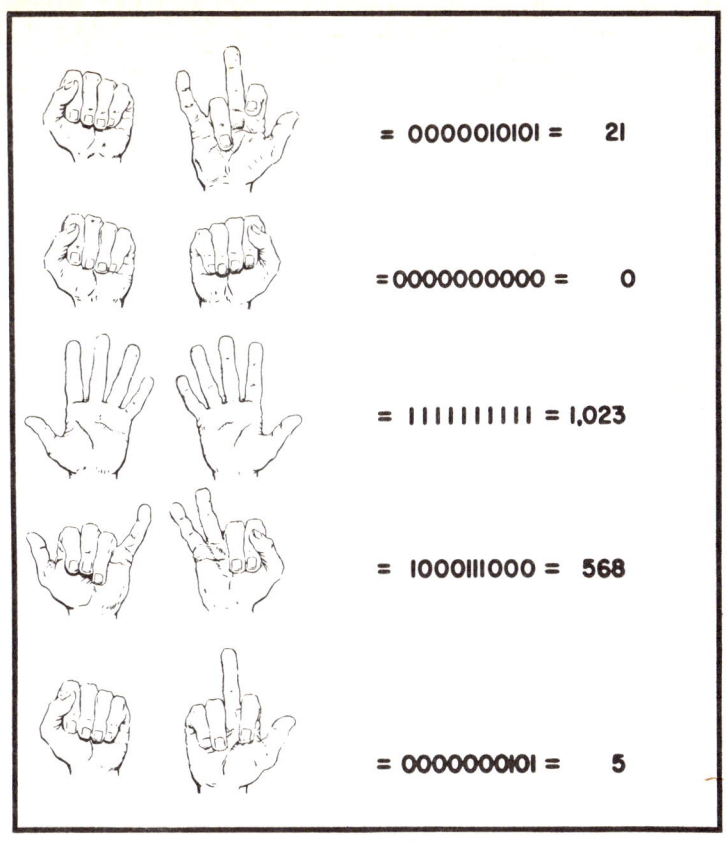

= 0000010101 = 21

= 0000000000 = 0

= 1111111111 = 1,023

= 1000111000 = 568

= 0000000101 = 5

Fig. 4-4. Finding the value of the binary number.

thus

$$0000010101 = 21$$

The values of the other binary numbers in the table are determined in the same manner. Just for drill, check them out on your fingers. Figure 4-4 will give you some answers.

Here's another way of considering the binary code. As you can see in Fig. 4-3, each digit (1 or 0) in a binary number has a different weighting value above it. We can multiply each digit by its weighting value, add all the products, and note the result. Using 0000010101 again, we obtain: $(0 \times 256) + (0 \times 128) + (0 \times 64) + (0 \times 32) + (1 \times 16) + (0 \times 8) + (1 \times 4) + (0 \times 2) + (1 \times 1) = 16 + 4 + 1 = 21$. This is the same value that we obtained before.

One more point must be noted. By convention, weighting values are always arranged in the same manner. The highest one is on the extreme left, and the lowest one is on the extreme right. Thus, position weighting values begin at 1 and increase from right to left. This convention possesses two very practical advantages.

First, it enables us to eliminate the table. We do not have to label each binary number with weighting values, because we know that the digit at the extreme right is always multiplied by 1, the next number to its left is always multiplied by 2, etc.

Second, we can often eliminate some of the zeros. Notice that the zeros, whether they are to the left or the right, never add to the value of the binary number. Zero times any number is zero. Thus, for practical purposes, only the 1s have to be multiplied by their weighting values and added to each other.

Some zeros are required, however. The zeros to the right of the highest valued 1 serve as place-keepers, or spacers, to retain the 1s in their correct positions. The zeros to the left, however, provide no information about the number and so they can be eliminated. Thus, 0000010101 can be written as 10101.

Significant Digits

The correct name for the left-most 1 in any binary number is the *most significant digit*. This is often abbreviated as MSD. It is the "most significant" because it is multiplied by the highest weighting value. The *least significant digit* (LSD) is, as you have probably guessed, the extreme right digit. Unlike the MSD, it may be a 1 or a 0. The LSD has the lowest weighting value: namely, 1. Figure 4-5 illustrates the meaning of MSD and LSD. These terms possess the same meanings in both the binary and decimal systems.

Converting From Decimal To Binary

Obviously, it is necessary to convert (by hand or mechanically) from the decimal to the binary system when making an input to a binary digital computer. It is also necessary to convert binary results to decimal form to produce a usable output.

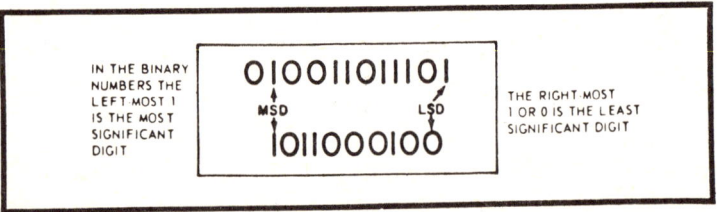

Fig. 4-5. The most significant digit, MSD, and least significant digit, LSD.

Our decimal system of Hindu-Arabic characters is to the base 10. There are 10 digits in this system. Each place in a decimal number represents a power of 10, starting with 10^0. (Any number to the zero power is equal to 1.) The digits give the frequency of the power of 10 for the particular place. For example:

$(4208)_{10}$ ← indicates number is to the base 10.

$$
\begin{aligned}
8 &\times 10^0 = 8 \times 1 &=& \quad 8 \\
0 &\times 10^1 = 0 \times 10 &=& \quad 00 \\
2 &\times 10^2 = 2 \times 100 &=& \quad 200 \\
4 &\times 10^3 = 4 \times 1000 &=& \underline{4000} \\
& & & (4208)_{10} \text{ or } 4208
\end{aligned}
$$

You just learned that the binary system is to the base 2, and that there are only two digits in this system. Each place in a binary number represents a power of 2, starting with 2^0. For example:

MSD LSD

$(101011)_2$ ← indicates number is to the base 2.

$$
\begin{aligned}
1 &\times 2^0 = 1 \times 1 = 1 \\
1 &\times 2^1 = 1 \times 2 = 2 \\
0 &\times 2^2 = 0 \times 4 = 0 \\
1 &\times 2^3 = 1 \times 8 = 8 \\
0 &\times 2^4 = 0 \times 16 = 0 \\
1 &\times 2^5 = 1 \times 32 = \underline{32} \\
& \qquad\qquad\qquad (43)_{10} \text{ or } 43
\end{aligned}
$$

In the previous example, besides showing you how each place in a binary number represents a power of 2, we also converted a binary number to its decimal equivalent. Now let's convert a decimal number to a binary number. We will convert the decimal number 173 to its binary equivalent:

$$
\begin{aligned}
(173)_{10} &= 128 + 45 \\
&= 2^7 + 32 + 13 \\
&= 2^7 + 2^5 + 8 + 5 \\
&= 2^7 + 2^5 + 2^3 + 4 + 1 \\
&= 2^7 + 2^5 + 2^3 + 2^2 + 2^0 \\
&= (10101101)_2
\end{aligned}
$$

This process of converting a decimal number to a binary number is known as the *subtraction method*. First find the highest power of 2 that can be subtracted from 173. In this case it is 2^7, or 128. You successively subtract the highest power of 2 that can be taken from the remainder. The powers of 2 that can be subtracted are noted by a 1. Zeros occupy the spaces not used.

Binary Addition

Binary digital computers can add, subtract, divide, and multiply. In the interest of brevity we will discuss only binary addition. Now let's consider the following pencil-and-paper addition of two binary numbers.

$$1 \text{ carried}$$

$$\begin{array}{r} 10 = 2 \\ \underline{11} = \underline{3} \\ 101 = 5 \end{array}$$

This addition is easily accomplished by observing these simple and logical rules:

$$0 \text{ and } 0 = 0$$
$$0 \text{ and } 1 = 1$$
$$1 \text{ and } 0 = 1$$
$$1 \text{ and } 1 = 0 \text{ and carry } 1$$

In the above example we first add the two digits in the first place column to get the result of 1. Then we add the two digits in the second place column, using the last rule listed, with a result of 0 and carry 1 to the next column on the left.

Here's another more complex example:

$$1 \quad \text{carried}$$

$$\begin{array}{r} 1\,1\,1\,1 \\ \hline 1\,0\,1 = 5 \\ 1\,1\,1 = 7 \\ \underline{1\,1\,1\,1} = \underline{15} \\ 1\,1\,0\,1\,1 = 27 \end{array}$$

Study these examples and make up a few of your own. Make sure you understand the binary addition process and then try converting each of the binary numbers to decimal form, adding them in a decimal column with the binary addition as proof.

Each of the binary numbers shown in Table 4-1 is called a *binary-coded decimal digit*, or *coded digit*. One can see that, without using any other binary numbers, it is possible to represent any desired quantity by means of these coded digits. For example, to represent 736, we simply write the coded digit

Table 4-1. Binary-coded decimal digits.

Decimal Digit	Binary - Coded Decimal Digit			
0	0	0	0	0
1	0	0	0	1
2	0	0	1	0
3	0	0	1	1
4	0	1	0	0
5	0	1	0	1
6	0	1	1	0
7	0	1	1	1
8	1	0	0	0
9	1	0	0	1

for each decimal digit; then we place the digit in the same order as the decimals. Thus: 736 = 011100110110.

Binary-coded decimal (BCD) digits need not be spaced, because we know that each one is composed of four binary digits. For easier reading, however, a space usually will be placed between each coded digit, as in 0111 0011 0110. Note that this number is not pure binary; it is, instead, a binary-coded decimal number.

COMPUTER LOGIC

Digital computers are used to make logical decisions (but not to reason) about matters that can be logically decided. Some logical decisions are: when to perform an operation, what operation to perform, and how (of several ways) to perform it. Digital computers never reason why or think out an answer; they operate only on instructions prepared by a man who has already applied the thought process to a problem and has reduced the problem to a point where logical decisions can deliver the correct answer. The rules for the equation and manipulations employed by the computer do not differ in many cases from the familiar rules and procedures of everyday mathematics.

People use many logical truths in everyday life without realizing it. Most of the simple logical patterns are distinguished by words such as *and*, *or*, *not*, *if*, *else*, and *then*. Once a verbal reasoning process has been completed and results put into statements, the basic laws of logic can be used to evaluate the process. Although simple logic operations can be

Table 4-2. Coincidence Amplifier Truth Table.

Control grid input present	Suppressor grid input present	Output signal present
NO	NO	NO
NO	YES	NO
YES	NO	NO
YES	YES	YES

performed by manipulating verbal statements, the structure of more complex relationships can be usefully represented by the use of symbols. Thus, the operations are expressed by what is known as *symbolic logic*.

The symbolic logic operations utilized in digital computers are based on the investigations of George Boole, and the resulting algebraic system is called *Boolean algebra*. It is similar in some aspects to the algebra with which you are familiar; however, it follows different laws and rules.

Truth Tables

A *truth table* is a chart; it is used in connection with logic circuits to illustrate the states of the inputs and outputs under all possible signal conditions. It provides a ready reference for use in analyzing the operating theory of the circuit and isurall signal-flow diagram.

In devising any truth table, it is necessary to know the number of inputs and the polarity of each. All possible states of the input are listed in column form, with a seperate column for each input line. The output for each combination of possible input states is then determined and noted in an output column. In logic circuits employing bistable devices, the number of possible input combinations in the truth table will be 2^n, where $n =$ number of input lines.

For example, a truth table constructed for a conventional pentode coincidence amplifier might be as shown in Table 4-2; the circuit is constructed in such a manner that an output occurs only when an input is applied simultaneously to the control grid and suppressor grid.

In constructing Table 4-2, all possible input combinations were placed in column form, and the output for each combination was determined and noted. In this particular example the number of possible inputs also follows the rule for bistable devices; that is, with two inputs the table contains 2^2, or 4, combinations.

Table 4-3. Truth Table Symbols for Coincidence Amplifier.

A	B	F
0	0	0
0	1	0
1	0	0
1	1	1

In computer logic truth tables, the column headings normally contain letter designations for the input and output, while the *yes* and *no* are replaced by symbols to denote the state of the inputs and output. The symbols include binary 1s and 0s, plus and minus signs, or *H* and *L* (*high* and *low*). Using symbols to construct the truth table for the coincidence amplifier would result in Table 4-3.

The meanings of the symbols in Table 4-3 are as follows:

$$A = \text{control grid input}$$
$$B = \text{suppressor grid input}$$
$$F = \text{output signal}$$
$$1 = \text{signal present}$$
$$0 = \text{no signal present}$$

When plotting a truth table for a circuit involving a large number of inputs, a convenient method of insuring that all possible inputs are listed is to start with all 0s and then count in binary sequence until the correct number of combinations has been listed. The last binary number listed in the input side of the table will be one less than the number of required inputs. Note in Table 4-3 that there are four input combinations, and the last input is 11 (say "one, one"), or binary 3.

Boolean Algebra

The objective of using Boolean algebra in digital computer study is to determine the *truth value* of the combination of two or more statements. As Boolean algebra is based upon elements having two possible stable states, it becomes very useful in representing switching circuits. The reason for this is that a switching circuit can be in only one of two possible stable states at any given time; that is, the state of being open or the state of being closed. These two states may be represented as *0* and *1*; we can use these symbols with the Boolean algebra.

In the mathematics with which you are familiar there are four basic operations—addition, subtraction, multiplication, and division. In Boolean algebra there are three basic operations—AND, OR, and NOT. If these words do not sound

Table 4-4. Logic Symbols.

Operation	Meaning
A · B	A AND B
A + B	A OR B
Ā	A NOT or NOT A

mathematical, it is only because logic began with words, and not until much later was it translated into mathematical terms. The basic operations are represented in logical equations by the symbols in Table 4-4.

The OR operation is indicated by the addition symbol, while the AND operation is indicated by the multiplication symbol. In addition to the dot, parentheses and other multiplication signs may be used. The negation function may also be indicated by A' instead of Ā. Examples of combinations of these symbols are given in Table 4-5.

Logic Operations

The three basic logic operations (AND, OR, and NOT) and four of the simpler combinations of the three (NOR, NAND, inhibit, and exclusive-OR) are shown in Fig. 4-6. For each operation, a representative switching circuit, a truth table, and a block diagram are given. In some instances, more than one variation is shown in order to illustrate some specific point in the discussion of a particular operation. In all cases, a 1 at the input indicates the presence of a signal (corresponding to switch closed) and a 0 represents the absence of a signal (switch open). In all outputs, a 1 represents a signal across the load, a 0 indicates no signal.

For the AND operation, every input line must have a signal present in order to produce an output. For the OR operation, an output is produced whenever a signal is present at any input. In order to produce a no-output condition, every input must be in a no-signal state. In the NOT operation, an input state produces an output, while a no-signal input state produces an output signal. (Note the block diagrams representing the NOT circuit in the figure.) The triangle is the symbol for an amplifier, and the small circle is the symbol for the NOT function. The circle is used to indicate the low-level side of the inversion circuit.

Table 4-5. Combinations of Logic Symbols.

Operations	Meaning
(A + B)(C)	A OR B, AND C
AB + C	A AND B, OR C
Ā · B	A NOT, AND B
A + B̄	A OR NOT B

The NOR operation is simply a combination of an OR operation and a NOT operation. In the truth table, the OR operation output is indicated between the input and output columns. The switching circuit and the block diagram also indicate the OR operation. The NAND operation is also a combined operation, comprising an AND and a NOT operation.

The inhibit operation is also a combination of an AND and a NOR operation, but the NOT operation is placed in one of the input legs. (In the example shown, the inversion occurs in the B input leg, but in actual use it could occur in any leg of the AND gate.)

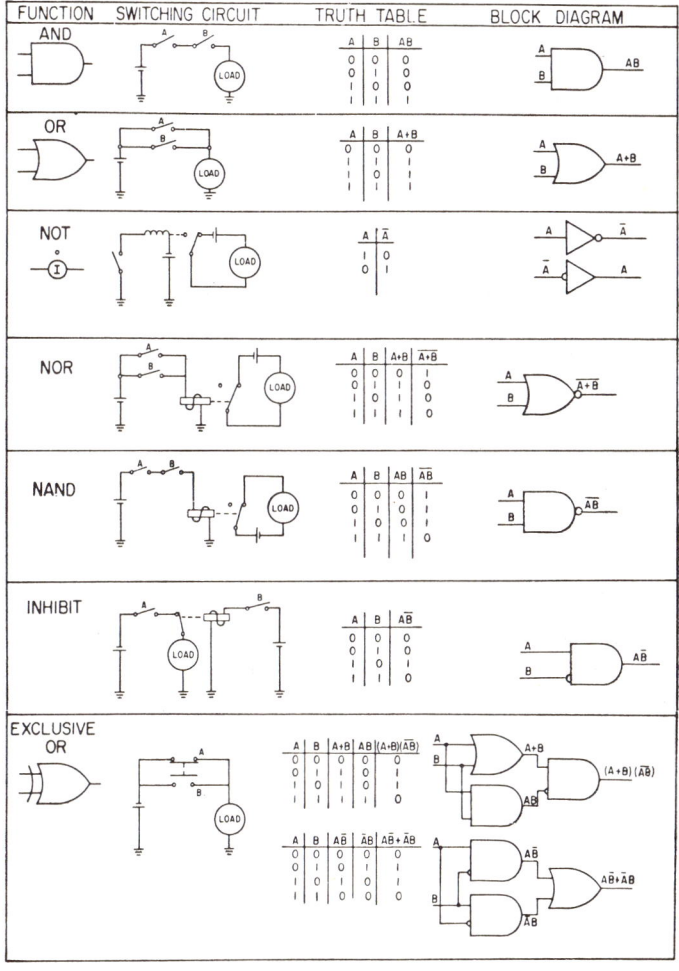

Fig. 4-6. Logic operations—comparison chart.

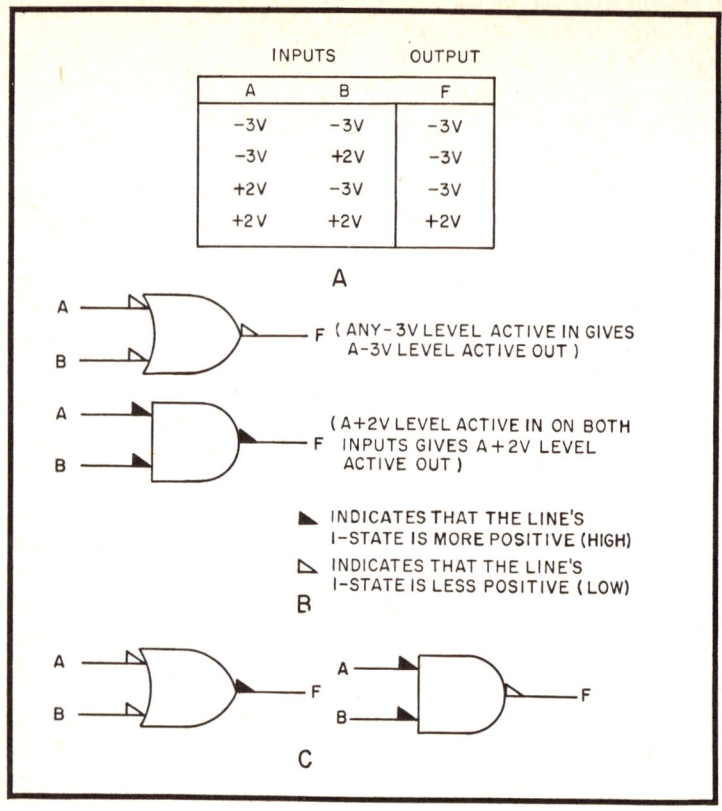

INPUTS		OUTPUT
A	B	F
−3V	−3V	−3V
−3V	+2V	−3V
+2V	−3V	−3V
+2 V	+2V	+2V

A

A ──▷ ──┐
B ──▷ ──┘ F (ANY−3V LEVEL ACTIVE IN GIVES A−3V LEVEL ACTIVE OUT)

A ──▶ ──┐
B ──▶ ──┘ F (A+2V LEVEL ACTIVE IN ON BOTH INPUTS GIVES A+2V LEVEL ACTIVE OUT)

▶ INDICATES THAT THE LINE'S I−STATE IS MORE POSITIVE (HIGH)

▷ INDICATES THAT THE LINE'S I−STATE IS LESS POSITIVE (LOW)

B

A ──▷ ──┐
B ──▷ ──┘ F

A ──▶ ──┐
B ──▶ ──┘ F

C

Fig. 4-7. Logic functions with high- and low-level inputs.

The exclusive-OR operation differs slightly from the OR operation. In the case where a signal is present at *every* input terminal, the OR produces an output and the exclusive-OR does not. In the switching circuit shown, both switches cannot be closed at the same time, but in actual computer circuitry this may not be the case. In the accompanying truth tables and block diagrams, two possible circuit configurations are indicated. In each case, the same final results are obtained, but by different methods.

Positive and Negative Logic

Logic circuits may be utilized to provide a "yes or no" or "true or false" answer to specific questions. They may be gates or switching circuits which are either conducting or cut off, or they may be circuits that will always be at one of two specific voltage levels. In either case, the two outputs may be considered as being either negative or positive with respect to each other. Normally, these output levels are used to represent binary 1s or binary 0s. As far as circuitry is concerned, it is possible to use either voltage state (or level) to represent either binary digit; therefore, it is essential to the understanding of circuitry that some distinction be made regarding the system of logic in any specific application.

The term *positive logic* is used to denote circuitry in which the voltage level utilized to represent a binary 1 is positive with respect to the level which represents a binary 0. The term *negative logic* is applied to circuits in which the voltage level representing a binary 1 is negative with respect to the level which represents binary 0.

Consider a circuit whose output is a function of two variables and whose input and output levels are capable of assuming only +2V and −3V as shown in Fig. 4-7A. This circuit can perform either the AND operation or the OR operation as shown in Fig. 4-7B.

In addition, it is possible to have a gate whose active output level is opposite to the activating input level, as shown in Fig. 4-7C. Note that Fig. 4-7C is the NOR and NAND functions discussed earlier. A logic of this type that permits the active function to be described in terms of either possible level (*high* or *low*), is referred to as *mixed logic*.

BASIC CIRCUITS

Many of the basic circuits used in digital computers are conventional electronic circuits generally applicable to electronic equipment. These circuits include amplifiers, timing-signal generators, pulse generators, waveshaping circuits, and time delay circuits. The circuits discussed here are limited to those logic circuits which are peculiar to digital computers.

Logic circuits are the high-speed electronic switching circuits that perform the functions of strong information signals and of performing logic operations on these signals. The circuits which perform the logic operations sense the input conditions and provide an output only if certain input conditions exist. They can be classified generally as gate circuits, and they may employ electron tubes, transistors, semiconductor diodes, or magnetic cores, as stated previously.

NOT Circuits

The NOT circuit, also known as an *inverter*, is nothing more than a normal amplifier stage that produces an inversion

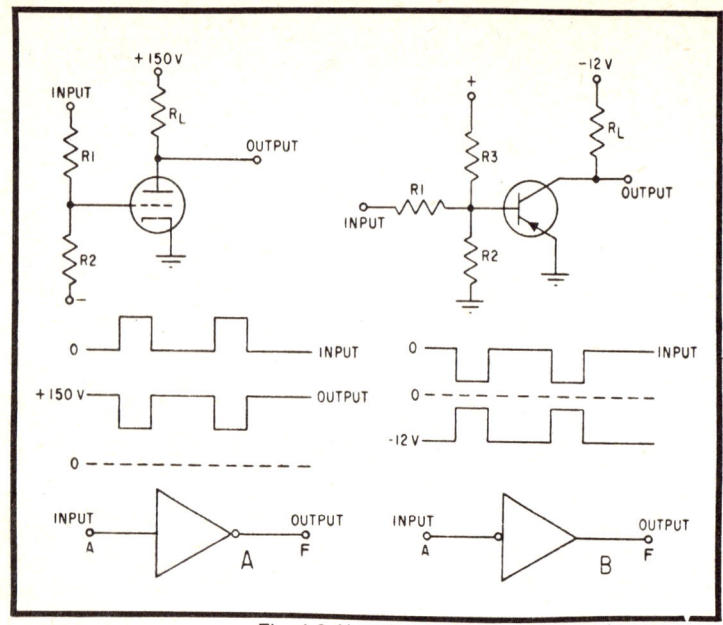

Fig. 4-8. NOT circuits.

(negation) of the input logic signal. Figure 4-8 illustrates two NOT circuits. The waveforms and logic diagram symbols for these two circuits are given.

In the electron tube circuit shown in (A), the triode is biased beyond cutoff with no signal applied. A positive pulse input causes the tube to conduct, and the output voltage level changes to a relatively negative value. (An electron-tube NOT circuit, using negative logic in the same configuration, would be biased for heavy conduction when no signal is applied, and negative inputs would cut the tube off.)

In the PNP transistor circuit shown in (B), the transistor is biased to cutoff by the voltage divider (R2 and R3), making the base positive with respect to the emitter. The input signal voltage is divided by R1 and R2 to make the circuit's output swing between the logic high and low voltage levels. Note that the input signal must be negative to cause conduction; therefore, this circuit represents a negative-logic NOT circuit. The logic symbols presented here indicate, by the use of a small circle on the triangle, the low level or negative state; that is, in (A) the inversion is from a positive-going to a negative-going signal, while in (B) the reverse is true.

It should be noted at this time that any circuit which inverts the signal falls under the definition of a NOT circuit.

OR Circuits

An OR circuit is a logic circuit having two or more inputs and producing an output if any one or more (or all) of the inputs are placed in the required signal state. Figure 4-9 shows two types of OR circuits.

In Fig. 4-9, four inputs are shown although this is not necessary; many OR circuits have only two inputs, but any practical number of inputs may be used. The truth table in this figure also illustrates that a two-level gate with 4 inputs will have 16 different input signal combinations ($2^4 = 16$). The waveform drawing does not show all possible input combinations, but enough combinations are shown to convey the idea that with any positive input, a positive output will result.

The electron tube in the circuit, shown in Fig. 10A, is biased to cutoff when there is no signal applied. The inputs are each fed through a resistor to the grid of the tube. The resistor in the signal path decreases the interaction between the individual input lines. A positive signal applied to any one or more of the input leads causes the tube to conduct, and (by cathode-follower action) a positive signal is taken from the cathode.

The positive-logic diode OR circuit (Fig. 4-10B) can be examined using the truth table in Fig. 4-9. When any one or more of the inputs receive a positive signal, the diode receiving that input conducts. A signal is impressed across R_L, resulting in a positive-going waveform at the output.

NOR Circuits

Fig. 4-11 shows a transistor-type OR circuit which also includes the negation function of the NOT circuit. A circuit that

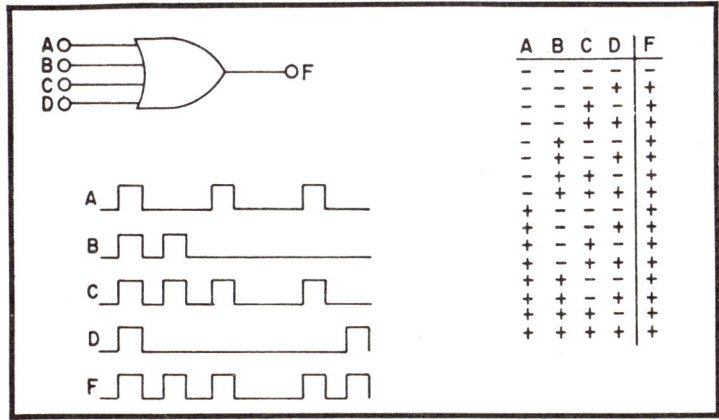

A	B	C	D	F
−	−	−	−	−
−	−	−	+	+
−	−	+	−	+
−	−	+	+	+
−	+	−	−	+
−	+	−	+	+
−	+	+	−	+
−	+	+	+	+
+	−	−	−	+
+	−	−	+	+
+	−	+	−	+
+	−	+	+	+
+	+	−	−	+
+	+	−	+	+
+	+	+	−	+
+	+	+	+	+

Fig. 4-9. OR circuit with four inputs.

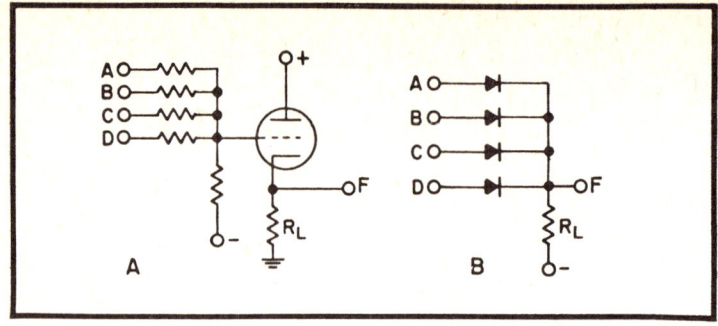

Fig. 4-10. Triode and diode OR circuits.

performs both of these logic functions is called a NOR (NOT OR) circuit, as explained previously.

With no input signal applied, the NPN transistor is biased to cutoff by the reverse bias between the emitter and base. A positive input pulse causes conduction of the transistor, producing a negative-going output pulse at the collector. The triode in Fig. 4-10A could be changed to a NOR circuit by replacing the cathode output with an output from a plate load resistor.

AND Circuits

An AND circuit is similar to a coincidence circuit in that it requires that all inputs meet a set of prescribed conditions in order to produce an output. Two transistorized versions of AND circuits are shown in Fig. 4-12.

In Fig. 4-12A, two PNP transistors are connected together at the collectors and emitters to form an emitter-follower circuit. There are two separate input terminals for the base signals.

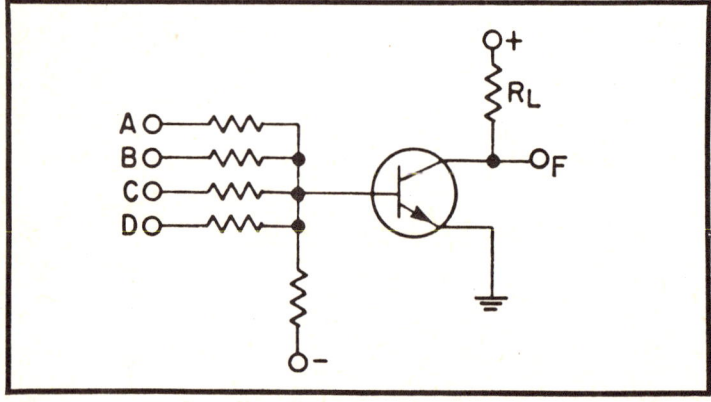

Fig. 4-11. Transistor NOR circuit.

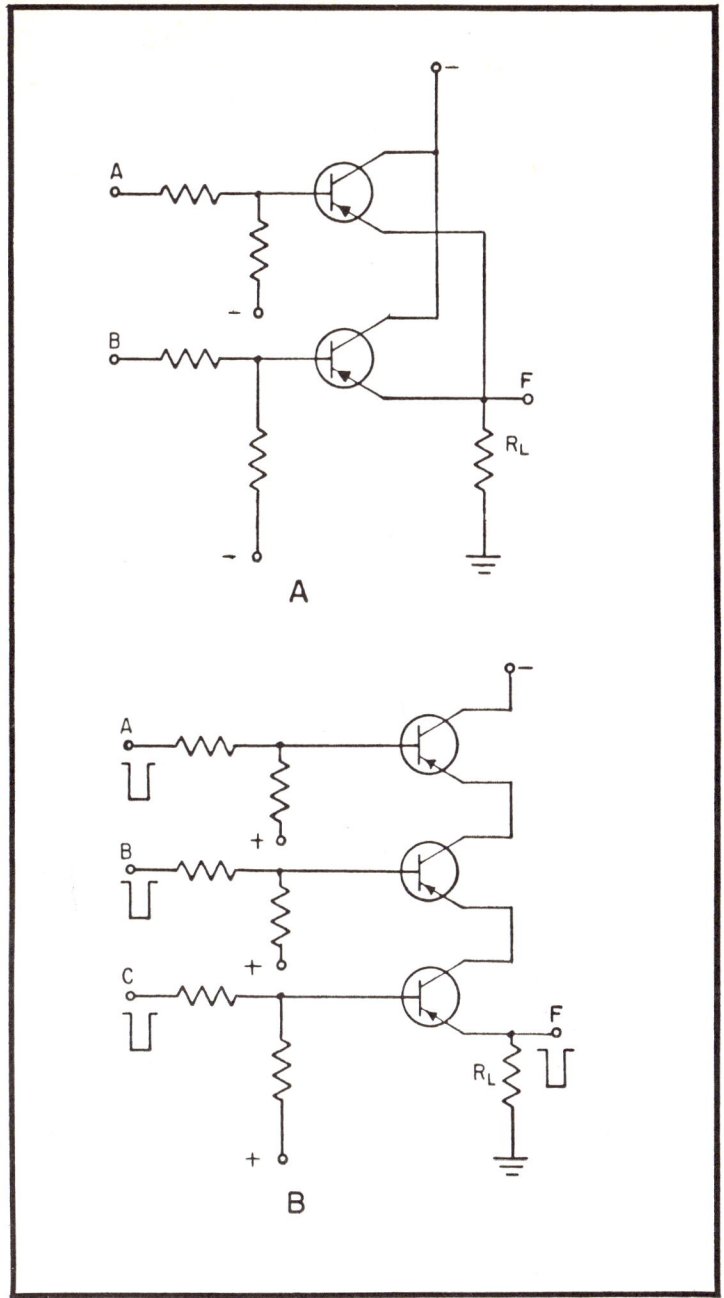

Fig. 4-12. AND circuits. (A) positive logic, (B) negative logic.

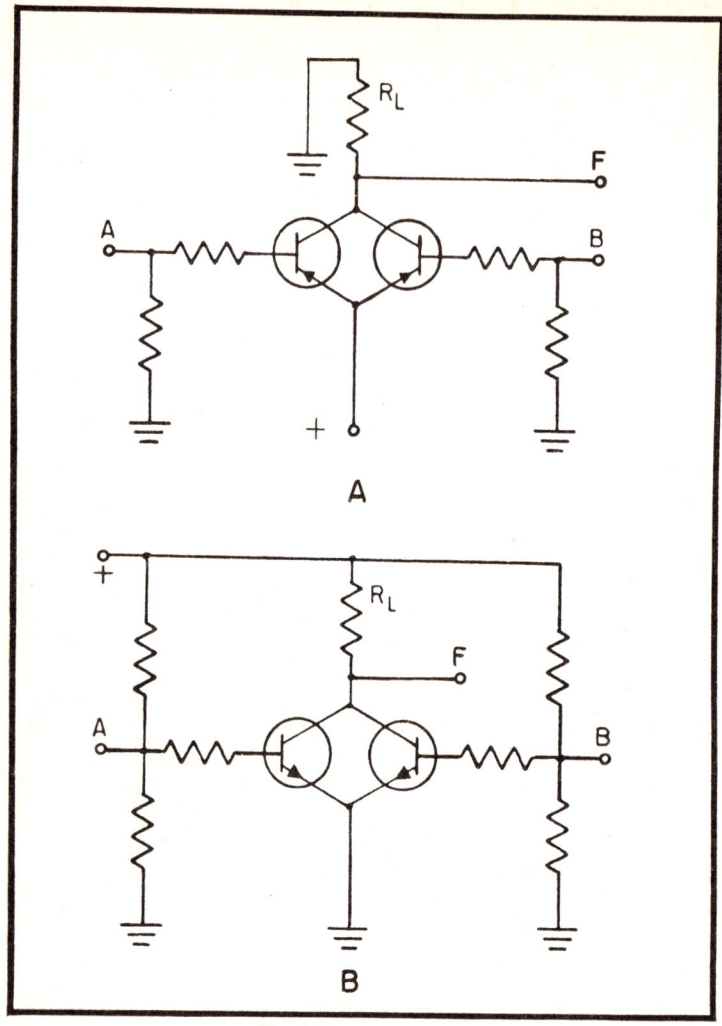

Fig. 4-13. NAND circuits.

Initially, both transistors are biased to saturation conduction, and nearly all of the collector voltage is dropped across R_L. Positive signals fed to the base of each transistor will cut off the respective transistors, but both transistors must be cut off in order to produce an output. When both of the transistors are cut off simultaneously, the output terminal voltage will go in a positive direction.

The circuit shown in Fig. 4-12B is a negative-logic AND circuit. With no signal applied, each transistor is cut off. During operation, all transistors must be activated together. If any one or more of the transistors remains cut off, there is no current path through the series-connected emitters and collectors. If all transistors are driven into conduction (by negative signals applied to the bases), the output at the emitter terminal will increase in a negative direction.

NAND Circuits

Two NAND circuits are shown in Fig. 4-13. Both of these circuits combine the AND function and the NOT function.

In Fig. 4-13A, the PNP transistors are biased for saturation conduction in the absence of signals. If positive signals are applied simultaneously to both bases, both transistors will be cut off (the AND function). When cutoff occurs for both transistors, the collector voltage (output) decreases negatively. This signal inversion is the NOT function.

In Fig. 4-13B, the NPN transistors are also biased for saturation conduction. When both transistors are cut off by coincident negative signals, the output voltage from the collector increases positively. This is a negative-logic NAND circuit.

Inhibitor Circuits

An inhibitor may be formed by using a combination of NOT and AND circuits, with the NOT circuit in one input leg of the AND circuit. This operation can also be performed by other circuits. The circuit in Fig. 4-14A provides the inversion (NOT function) by placing a transformer in the input leg of diode D3.

Diodes D1 and D2 are forward biased, allowing them to conduct. This places the output voltage at a low level (binary 0). Diode D3, however, is biased so that the diode is cut off with no signal applied to input C. With a signal applied to either input A or B, either D1 or D2 will be cut off; however, the other will continue to conduct, holding the output at a low potential.

When inputs A and B both have a signal applied, with no signal applied to input C, both D1 and D2 will be cut off and the output will rise to a high positive potential (binary 1).

If a pulse is applied to input C, the pulse will be inverted across the transformer. This will overcome the bias on D3, allowing it to conduct, holding the output to a low level regardless of the state of the other inputs.

Another circuit which will perform the inhibit operation is shown in Fig. 4-14B. This circuit employs an NPN transistor with the emitter grounded. Note that the emitter−base circuit is

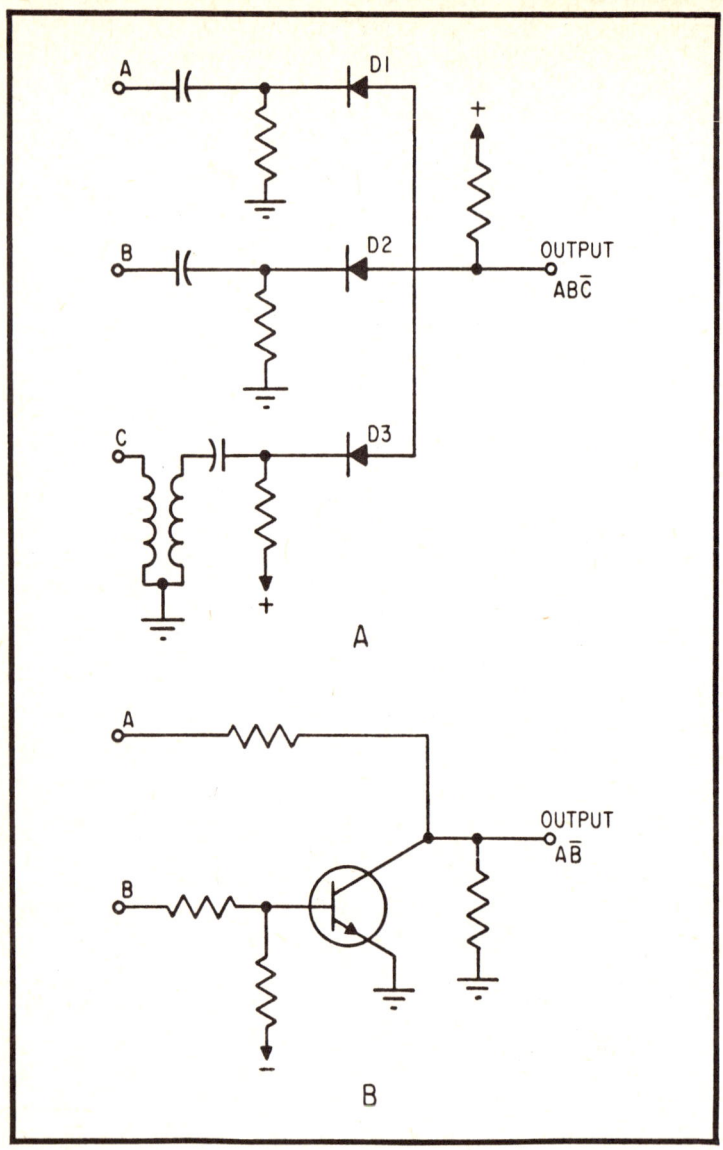

Fig. 4-14. Inhibitor circuits.

biased in the reverse direction; therefore, the transistor will be normally cut off.

If a positive pulse is applied to input *A*, current flow through the load resistor will produce a corresponding signal at the output terminal, when an input is present as input terminal *B*.

106

If, however, a positive pulse appears at input *B* at the same time there is one present at input *A*, this input would overcome the reverse bias and the transistor would conduct to saturation. When at saturation, the transistor is effectively a short circuit and the input pulse at *A* would be effectively lost; thus it would not appear at the output. In this fashion, a pulse applied to input *B* inhibits the appearance of a pulse at the output terminal.

Flip-flop Circuits

The flip-flop circuit (bistable multivibrator) is used in digital computers to supply an output and its complement simultaneously, as an off/on trigger, and for storage purposes. It forms the basic circuit used in most registers.

Like other logic symbols, the flip-flop logic symbol is functional; it is the same regardless of internal circuit. Thus, any number of multivibrator types may be represented by the same logic symbol. One flip-flop symbol is a rectangle (Fig. 4-15A) with terminals extending from opposite sides to indicate *set* and *reset* input terminals and *set* and *reset* output terminals. (*Reset* is used interchangeably with *clear*.) The OR circuits and inverter contained in the flip-flop are not shown in the symbol, although an understanding of the operation of these circuits in the flip-flop is essential in determining what effect the various types of input pulses will have on the output.

A second flip-flop symbol (Fig. 4-15B) uses a toggle (T) or trigger input terminal. An input signal on this terminal is applied to both halves of the flip-flop and causes both sides to change their conducting condition; i.e., from 0 to 1, or vice versa.

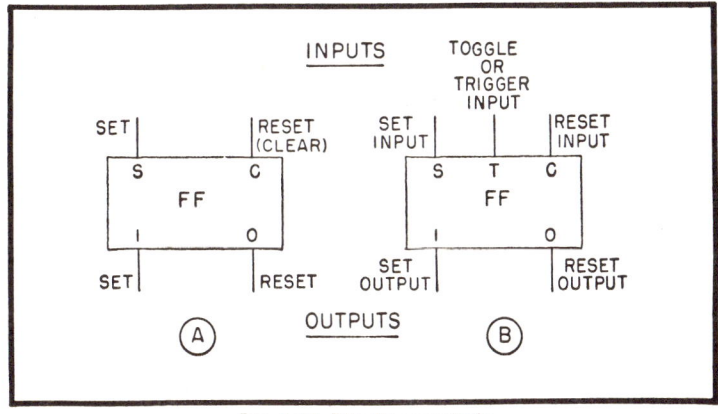

Fig. 4-15. Flip-flop symbols.

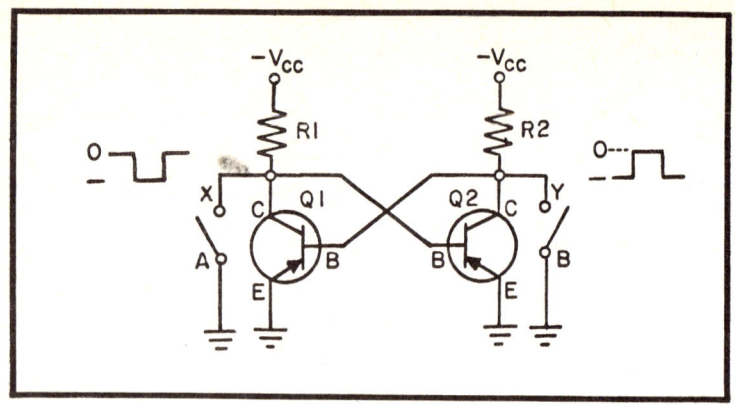

Fig. 4-16. Transistor flip-flop circuit.

The schematic of a simple transistor flip-flop is shown in Fig. 4-16. Outputs are taken from points X and Y, and inputs are represented by switches A and B which temporarily ground these points when the input is considered to be applied. The manner in which the trigger is applied is of no immediate concern since it may be in a number of ways, such as by diodes, transistors, relays, switches, or otherwise.

Assume that initially transistor Q2 is conducting, and Q1 is cut off. With Q2 operating in the saturation region, heavy collector current flows and the collector voltage is dropped to almost zero by current flow through R2, and the Y output is zero. As the collector of Q2 is connected to the base of Q1, the base of Q1 is also held at zero bias or practically at cut off, and no collector current flows through Q1. With no collector current flow through R1 the collector voltage on Q1 is almost the full value of the negative bias supply.

As Q2 is directly connected to the collector of Q1, this negative collector voltage places a large forward bias on the base of Q2 and holds it in heavy conduction. As long as the heavy collector current flows through Q2, transistor Q1 remains in the inactive state. Thus, the Y output is held at zero, representing a logic 0, while the X output is a large negative voltage, representing a 1.

If switch A is now temporarily closed (simulating a set input), the forward bias is removed from the base of Q2 because the collector of Q1 is shunted to ground; therefore, collector current ceases flowing through Q2. As the collector current through Q2 ceases, the collector voltage of Q2 rises toward the negative bias source value, and places a forward bias on the base of transistor Q1. Consequently, Q1 conducts heavily, and the voltage drop across collector resistor R1 places the base of Q2 at zero and prevents the flow of collector current in Q2. Now, Q2 remains in the nonconducting condition, while Q1 continues to conduct heavily.

Meanwhile, the Y output is a large negative voltage, representing a 1. When switch B is temporarily closed (simulating a clear input), the base of Q1 is grounded, and causes it to stop conducting. The rising collector voltage on Q1 places a negative (forward) bias on the base of Q2 and causes it to conduct heavily and resume the original state assumed at the beginning of the discussion. This action again produces a 0 output at Y, while the X output is a 1.

Thus, when Q2 conducts, the X output is 1 and the Y output is 0; and when Q1 conducts, the Y output is 1 and the X output is 0. The alternate X and Y outputs of 0 and 1 continue as the flip-flop is triggered off and on by closing switch contacts A and B. Although relay triggering is assumed in this discussion, actual computer circuits use switching diodes or transistors, which operate at much greater speeds.

COUNTING BY FLIP-FLOP

Assume that four flip-flop circuits are connected in cascade to form a four-digit binary counter, as in Fig. 4-17. Indicator lamps may be used to indicate a 1 or *set* condition when they are glowing; or a 0 or *reset* condition when they are off. The four indicators may be read from left to right as a four-digit binary number. The input pulses to be counted are applied to the right-hand flip-flop, which represents the least significant digit.

Assume that all four flip-flops are in the zero (reset) condition and a negative input appears at the input (Fig. 4-17A). Flip-flop FF_1 changes state (reset to set) and its indicator goes on; but FF_1 produces no output, and the other three flip-flops continue to indicate logic 0. The four indicators may be read as a binary number: 0001.

With the second input (not shown) FF_1 returns to 0, and provides a negative-going input to FF_2. Flip-flop FF_2 changes to the logic-1 state, but produces no output. The binary indication is now 0010 (the binary equivalent of decimal 2), indicating that two inputs have been counted.

With the third input (Fig. 4-17B), FF_1 changes to the logic-1 (set) state, but it provides no output. Flip-flop FF_2 therefore remains in the logic-1 state, and the binary indication is now 0011 (decimal 3).

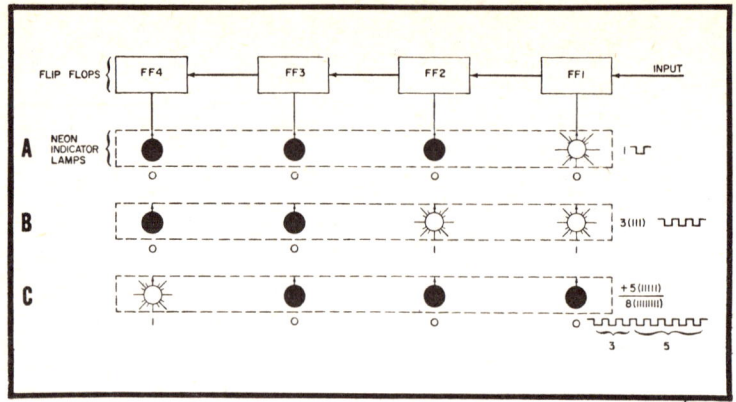

Fig. 4-17. Adding function of flip-flop.

Now let 5 more pulses go into the flip-flops. The first pulse of this group of 5 produces outputs from FF_1 and FF_2, since they are on. It also switches FF_3 on, but in going through, it turns FF_1 and FF_2 off. The second pulse of the 5 merely turns FF_1 on, so that the indication is 0101 which is 5 in binary notation. You can follow this same sequence (with the assistance of your fingers and your knowledge of the powers of two) up to 8, at which point the flip-flops will indicate 1000, as shown in Fig. 4-17C.

After 15 inputs, the lamps will indicate the binary number 1111 (decimal 15). On the sixteenth input, FF_1 changes to logic 0 and produces an output; FF_2 therefore changes to logic 0 and produces an output; FF_3 therefore changes to logic 0 and produces an output, which changes FF_4 to logic 0. The counter therefore returns to its original state (0000) after 16 inputs. If a fifth flip-flop were added, receiving its input from FF_4, the counter could count up to 11111 (decimal 31), and return to the original state on the 32nd input. And so on.

This example illustrates the counting function possible with a series of flip-flops. With modifications and elaborations of this, and the addition of other circuits, it is possible to perform all the fundamental arithmetic operations.

Note too that the flip-flops are capable of *storing* information—in other words, they have *memory*. In this case they remembered the binary work for 3 until the pulse train representing 5 came along. These two numbers were added and the binary sum was stored. In the digital computer, the unit which stores a number, then adds another number to the stored number, and then stores the sum of the two, is called an *accumulator* (discussed later).

ADDERS

Adders in digital computers use combinations of AND, OR, NOT, and inhibitor circuits to perform the functions of addition. When carries are not needed, a simple OR circuit could easily add numbers, for it has an output pulse whenever pulses appear at either or both inputs.

Binary addition is performed as follows when a carry is involved:

$$\begin{array}{r} 1 \\ 001 \\ \underline{101} \\ \overline{110} \end{array}$$

In the first column, a 1 is to be added to a 1; the sum is 0 with a 1 carry. The 1 carry is shown over the second column. Adding the 1 carry to the two 0s in the second column produces a sum of 1, with no carry. In the third column, a 1 added to a 0 produces a sum of 1. Thus, 110 is the correct sum.

One of the circuits that must be considered in connection with serial adders is the delay circuit. The delay circuit functions much like a conventional delay line. It delays a pulse train for the duration of one pulse; that is, all the pulses in a train advancing from left to right are moved one place to the left. Thus, the binary number 001 applied to a delay circuit would become 010 in its output. The delay circuit permits the carry function of arithmetic in adders. It is used with a combination of OR, AND, and NOT circuits to make up a full-adder circuit.

A serial half-adder circuit is shown in Fig. 4-18A. The OR circuit produces an output when there is an input pulse applied to either the A or B input. The lower AND circuit produces an output when A and B input pulses are received simultaneously, a condition which produces a carry.

First consider the action with 010 at the A input terminal and 101 at the B terminal as shown. Although the inputs are applied simultaneously to the OR circuit and the lower AND circuit, only the OR circuit operates to produce an output. The two simultaneous 1 inputs, necessary to produce a 1 output from the lower AND circuit, do not occur in this example and thus no carry is produced. The output of the OR circuit (111) is therefore the true sum of the binary numbers appearing at A and B.

The inverter at one terminal of the second AND circuit (inhibitor, discussed earlier) causes this input to be in the logic-1 state when the lower AND circuit output is 0 (as in this

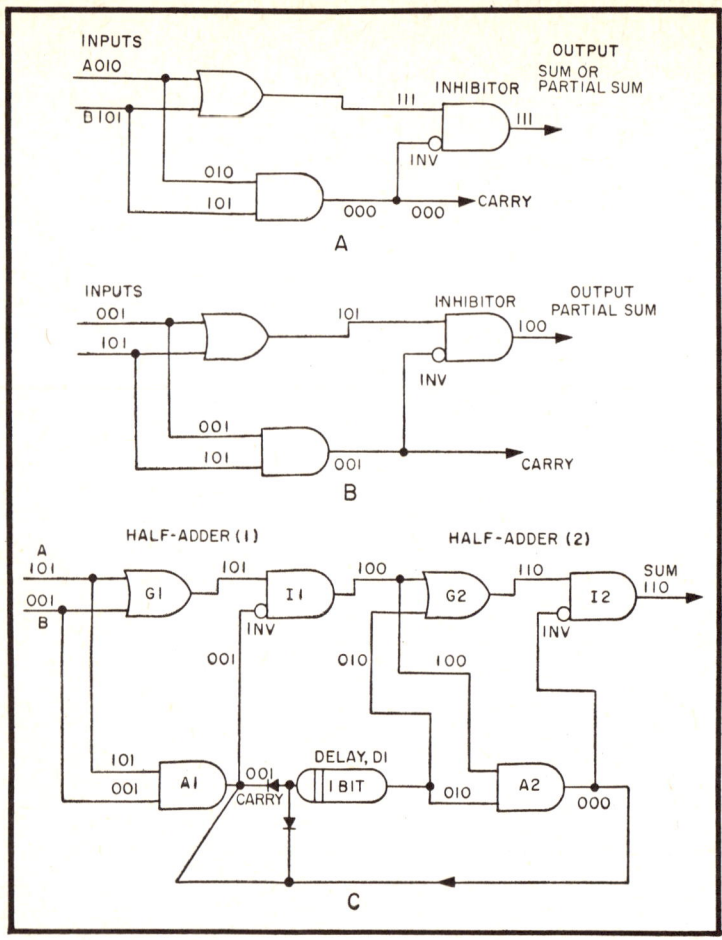

Fig. 4-18. Serial adders. (A) and (B), half adders. (C) full adder.

case). The 111 output from the OR circuit is then passed through the inhibitor to the half-adder output.

The addition process is not so simple when two logic-1 inputs occur simultaneously in the numbers to be added, as shown in Fig. 4-18B. Here, the OR circuit produces a 101 output (which is not a true sum of inputs A and B) while the AND circuit produces a 001 output. The 1 input to the inverter at the inhibitor input causes this circuit to block the passage of the 1 output from the OR circuit in the right (least significant) column, and the inhibitor output is 100. This, again, is not the true sum of the input binary numbers (001 and 101). Further, a true sum of these numbers cannot be obtained in a half-adder circuit.

In order to obtain the true sum, the carry generated by the AND circuit must be utilized in a subsequent half-adder circuit (2) as shown in Fig. 4-18C. In this circuit, the actions to produce the 100 output from inhibitor I1 and the 001 carry output from AND A1 are the same as described above. The 100 output of I1 is fed to the upper terminal of OR circuit G2. The carry output from A1 is delayed by D1 by one bit-time (a delay equal to the period of one binary digit). The 1 output from D1 will then be applied to the lower terminal of G2 after being displaced one position to the next higher-order column. This action advances the carry and thus produces a 110 output from G2.

Because there is no further carry generated at the output of A2, inhibitor I2 passes the number 110 to the sum output terminal. This is the true sum of the A and B inputs. The circuit is called a *full-adder* because it is able to produce a true sum of two (or more) inputs by generating and utilizing the carry when two 1 inputs occur simultaneously.

ASTABLE MULTIVIBRATOR

The transistor *astable* (free-running) multivibrator produces a square-wave output for use as a trigger or timing pulse in electronic equipment. The basic circuit, when modified for application to switching circuitry, is representative of a class of circuits which perform computer operations (counting, shift register, and memory circuits), control functions (relay driver circuits), and a variety of similar applications.

Forward bias for the base of transistor Q1 (Fig. 4-19) is obtained through the low-resistance emitter−base junction, which is in series with resistor R2 across the voltage source, V_{CC}. In a like manner, forward bias for the base of transistor Q2 is obtained through the emitter−base junction and resistor R3. When voltage is first applied to the multivibrator, the current which flows in each collector load resistor (R1 and R4) is determined by the effective resistance offered by transistors Q1 and Q2 for a given value of base-bias voltage.

For the purpose of this explanation, assume that initially more collector current flows through transistor Q1 than through transistor Q2; thus, as the collector current of Q1 increases, the voltage at the collector of Q1 decreases with respect to its emitter, or ground. In other words, the collector of Q1 becomes less negative and this, in effect, acts as a positive-going pulse which is coupled through capacitor C1 to the base of transistor Q2.

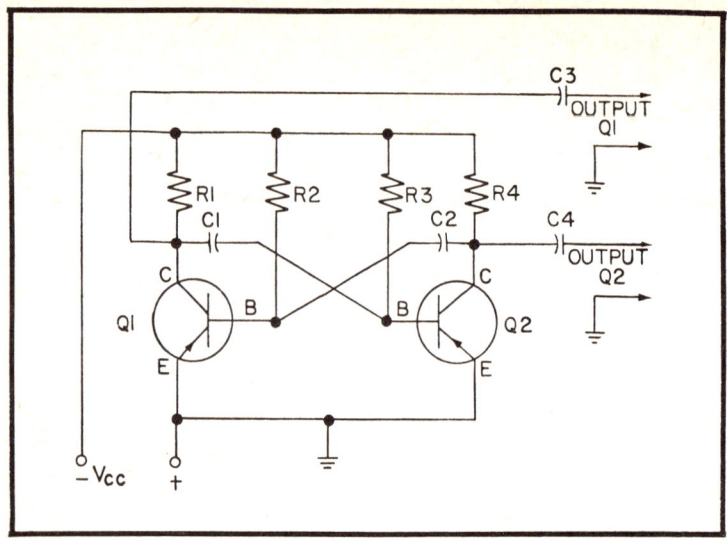

Fig. 4-19. Basic transistor astable multivibrator.

The positive-going pulse at the base of Q2 makes the base positive with respect to its emitter and, as a result, Q2 approaches cutoff. The collector current of Q2 increases, and approaches the supply voltage, V_{CC}. Thus, the collector of Q2 becomes more negative and this, in effect, acts as a negative-going pulse, which is coupled through capacitor C2 to the base of transistor Q1.

The negative-going pulse at the base of Q1 places the base negative with respect to its emitter, and the collector current of Q1 is further increased because of the forward-bias action. This regenerative process continues until Q1 is driven into saturation (as a result of increased forward bias), and Q2 is cut off (as a result of the reverse bias).

When Q1 is at saturation, its collector current no longer increases but becomes a constant value; therefore, there is no further change in collector voltage to be coupled through capacitor C1 to the base of transistor Q2. The voltage at the base of Q1 is only a few tenths of a volt negative; as a result, capacitor C2 quickly charges through the low resistance of R4 to a potential which is approximately equal to V_{CC}. Since the collector voltage at Q1 (Q1 is conducting heavily) is at nearly ground potential, capacitor C1 (previously charged) starts discharging (at a rate which is equal to the time constant R3 × C1) through transistor Q1, the voltage source, and resistor R3.

As capacitor C1 discharges, the voltage at the base of Q2 becomes less and less positive (negative-going) until a point is reached where reverse bias is no longer applied and Q2 is then able to conduct.

When the base of Q2 returns to a forward-bias condition, Q2 begins to conduct and its collector current begins to flow through load resistor R4. As the collector voltage at Q2 drops, a changing (positive-going) voltage is coupled through C2 to the base of transistor Q1. The voltage at the base of Q1 is only a few tenths of a volt negative; as a result of the charge on capacitor C2, reverse bias is applied to the base of Q1.

Transistor Q1 is thus driven to cutoff, and the collector voltage of Q1 rises. This rise, coupled through C1, will drive the base of Q2 further into the forward-bias condition. The voltage at the base of Q2 is only a few tenths of a volt negative, and the collector voltage at Q1 is approximately equal to V_{CC}. Since the collector voltage at Q2 (Q2 is conducting heavily) is at nearly ground potential, capacitor C2 (previously charged) starts discharging (at a rate which is equal to the time constant $R2 \times C2$) through transistor Q2, the voltage source, and resistor R2.

As capacitor C2 discharges the voltage at the base of Q1 becomes less and less positive (negative-going) until a point is reached where reverse bias is no longer applied, and Q1 is able to conduct.

When the base of Q1 begins to conduct, collector current begins to increase through load resistor R1. As the voltage drops at the collector or Q1 a changing (positive-going) voltage is coupled through capacitor C1 to the base of transistor Q2 to initiate another cycle of operation.

For each half-cycle of operation, whenever a change-over of the multivibrator takes place, one of two actions occurs: (1) capacitor C1 recharges through load resistor R1 and the base—emitter junction of Q2 to the value of the supply voltage, V_{CC}, while capacitor C2 discharges through the series circuit consisting of transistor Q2, the voltage source, and resistor R2; or (2) capacitor C2 recharges through load resistor R4 and the base—emitter junction of Q1 to the value of the supply voltage, V_{CC}, while capacitor C1 discharges through the series circuit consisting of transistor Q1, the voltage source, and resistor R3.

TRANSISTOR MONOSTABLE MULTIVIBRATOR

The transistor *monostable* (one-shot) multivibrator is a triggered circuit, which requires a trigger pulse to initiate

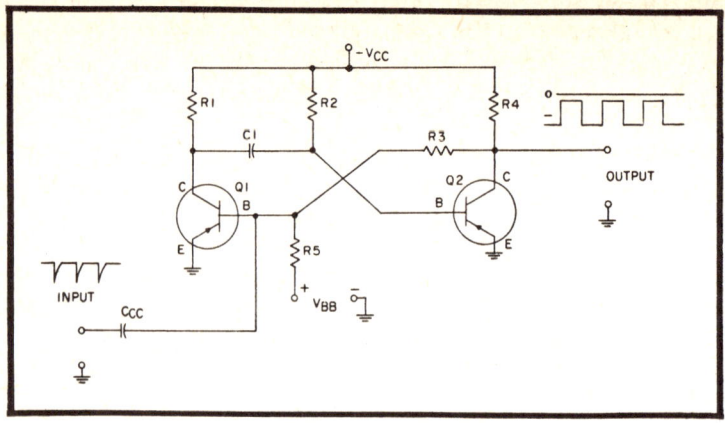

Fig. 4-20. Basic transistor monostable multivibrator.

action. Once the trigger pulse initiates the action, the circuit uses its own power to complete the operation. Either the stable state of cutoff or saturation is used. Normally, one transistor is operated saturated while the other is at cutoff. When the circuit is triggered by an external pulse, the operating point is moved from the initial stable region to the other stable (operating) region. Meanwhile, the time constant of the circuit elements holds the operating point in the new stable (operating) region for a short period of time. The operating point then moves back to the initial stable region.

Fixed forward bias is applied to the base of Q2 (Fig. 4-20) via resistor R2, while the voltage divider (consisting of R4, R3, and R5) forms a fixed bias divider between the base bias supply the collector supply and ground. Thus, Q1 is biased slightly positive and is cut off by this reverse bias. Resistor R4 also is the collector resistor for Q2, and R1 serves a similar function for Q1. Resistor R3 serves as the collector-to-base feedback resistor for Q1, while C1 is the feedback capacitor for Q2. Both emitters are grounded and a cross-connected, grounded-emitter circuit is used. The input is applied through coupling capacitor C_{cc}, while the output is taken directly from the collector of Q2. If desired, the output load could also be capacitively coupled.

In the quiescent condition, transistor Q2 conducts heavily while transistor Q1 is cut off. This action occurs initially because of the large negative forward bias placed on the base of Q2 via resistor R2, which is connected back to the negative supply. Thus, on application of power, Q2 quickly saturates and develops a positive-going output at the Q2 collector, which is fed back to the base of Q1 through resistor R3, holding transistor Q1

116

at cutoff. During the *on* period of Q2, the right side of feedback capacitor C1 is charged positively, through R1 and the low base—emitter saturation resistance of Q2. The low saturation resistance of the Q2 base—emitter junction acts as a switch, connecting R1 and C1 in series with the negative supply source and ground.

When the negative trigger is applied to the Q1 base through coupling capacitor C_{CC}, transistor Q1 is instantly driven into conduction by this forward bias. The flow of Q1 collector current through R1 reduces the effective collector voltage and produces a positive-going voltage at the Q1 collector, which is applied through feedback capacitor C1 as a positive reverse bias to cut off Q2.

As the collector current of Q2 decreases, the voltage across collector resistor R4 rises towards that of the negative collector supply, and an increasing forward bias is fed back to the base of Q1 through feedback resistor R3. Thus, Q2 is cut off and Q1 is turned on. Operation is now reversed and the output from Q2 is a negative voltage. Since C1 is positively charged, when disconnected from ground by Q2 being driven into cutoff, the capacitor holds the base of Q2 highly positive (reverse biased) while it discharges. The discharge path is through the low collector-to-emitter saturation resistance of Q1 to ground on one side, and through R2 to the negative supply on the other side.

Transistor Q2 remains nonconducting until the base voltage drops to zero. The base of Q2 then goes slightly negative and Q2 immediately starts to conduct. The flow of collector current through R4 produces a positive-going voltage, which is applied through feedback resistor R3 to drive Q1 in a reverse-biased direction and to stop conduction through Q1. This regenerative feedback action occurs quickly, and the output of Q2 is now a positive square-wave voltage. The quiescent state of operation continues until the next trigger, whereupon the switching action is again repeated.

The *squaring* multivibrator in Fig. 4-21 is also known as the Schmitt trigger or emitter-coupled bistable multivibrator. It is used primarily to supply a square or rectangular output when triggered by a sine-wave, sawtooth, or other irregularly shaped waveform.

The circuit differs from the conventional bistable multivibrator circuit in that one of the coupling (feedback) networks is replaced by a common-emitter resistor (the equivalent of cathode coupling in the electron tube). The additional regenerative feedback developed by the

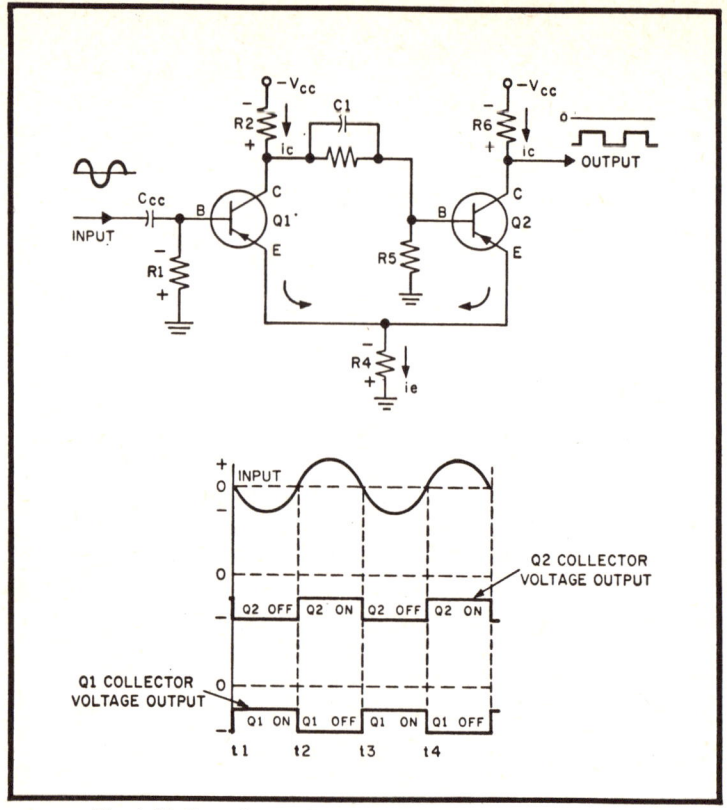

Fig. 4-21. Squaring multivibrator circuit and waveforms.

common-emitter feedback coupling arrangement provides quicker action as well as straighter leading and trailing edges on the output waveform than in other multivibrators. Because of the relatively fast switching action of this arrangement, the waveform of the input trigger has no effect on the output waveform, so that essentially square-wave output signals are always produced.

Transistor Q2 (Fig. 4-21) is the initially conducting transistor, which is supplied with forward base bias by resistor network R2, R3, and R5 connected as a voltage divider between the negative voltage supply and ground. Capacitor C_{cc} and base resistor R1 form a conventional RC-input coupling circuit. Resistor R4 is the feedback (coupling) resistor which is common to both emitters. Resistor R6 is the collector resistor of Q2, across which the output waveform is developed. Capacitor C1 bypasses feedback resistor R3 to speed up switching action.

Initially, transistor Q2 conducts heavily because of the large forward bias supplied by the voltage divider consisting of resistor R2, feedback resistor R3, and base resistor R5, series-connected between the negative supply and ground. A reverse bias is applied to the emitter of Q1 by the voltage developed across common-emitter resistor R4 by the current flow of Q2. With Q2 conducting, the output voltage developed from the collector to ground is approximately equal to the drop across R4.

Assume that a sine-wave input signal is applied to C_{CC}. During the positive half-cycle of operation, the positive input voltage applied across R1 keeps Q1 reverse biased so that it cannot conduct. When the input signal swings negative during the opposite half-cycle of operation (time t_1), a negative voltage appears at B as C_{CC} discharges through R1. The base of Q1 is thereby driven negative and forward biased, starting collector current flow through R2. The direction of electron flow is such that the collector end of R2 swings in a positive direction, and this instantaneous positive swing is coupled through C1 to the base of Q2, appearing as a positive reverse bias which instantly stops current flow through Q2.

The reduction of collector current flow through R6 causes a negative-going output voltage, as the collector of Q2 rises towards the negative supply voltage. Although R3 connects the collector of Q1 to the base of Q2, and any voltage appearing on the collector of Q1 will also eventually appear on the base of Q2, a speedup of this action is obtained by bypassing R3 with capacitor C1.

Consider now the effect of the common coupling resistor (R4) in the emitter circuit. Initially (prior to time t_1) the heavy current through Q2 produced a negative voltage at the Q2 emitter. This negative emitter bias, which is degenerative because no bypass capacitor is employed, also tends to prevent Q2 current flow. However, the base bias of Q2 is much more negative so that the degenerative emitter voltage produced across R4 has little effect on Q2 collector current flow. Transistor Q1 is already at cutoff (prior to time t_1) and this additional negative emitter bias insures that it remains off until sufficient base input is applied (on the next half-cycle) to overcome this reverse bias.

With the collector flow of Q2 decreasing, the degenerative voltage developed across R4 also reduces, which is the same as applying an increasing positive-going voltage between emitter and ground. Thus, while the base of Q1 is driven in a forward-biased direction by the input signal, a regenerative feedback is developed in the emitter circuit by the reduction of Q2 current flow. Consequently, transistor Q1 is quickly driven into heavy conduction. The resulting positive-going voltage is developed at the Q1 collector, applied instantly through C1 to the base of Q2, and quickly drives Q2 to cutoff. The circuit now rests in its second stable state (time t_1 to t_2) until another trigger arrives to drive Q2 into conduction and cut off Q1.

Chapter 5

Complex Logic Functions

Digital circuits may process information pulses in either serial or parallel form. In serial form, the pulses occur as a timed series, one pulse at a time (Fig. 5-1). In parallel operation, all pulses occur simultaneously and are transferred within the computer on separate transmission lines.

The counter circuit shown in Fig. 5-2 can be used to perform several computer operations. For example, consider a series of four toggle pulses at the input to the first flip-flop. The counter will advance to a state where the *set* outputs will indicate 0100, which may represent a computer word. Assuming all flip-flops are initially in the logic-0 state, each input pulse toggles FF_0; i.e., if FF_0 is in the 0 state, an input pulse changes it to the 1 state. Each subsequent flip-flop is toggled when the preceding one is toggled from the 1 to the 0 state.

The counter is a static device and will remain in the condition described above unless subsequent pulses are applied. The circuit thus "remembers" or stores this count. In this respect, a flip-flop counter may be used as a *register*, since it stores a computer word.

A second application of this circuit is seen when it is assumed that five additional pulses are applied to the counter input. The count now advances to 1001 (9) which is the sum of the two input-pulse series. This process is called serial addition. This circuit arrangement thus acts as a basic accumulator.

PARALLEL ADDER

The circuit shown in Fig. 5-3 is used to illustrate the basic example of parallel addition. Note that four binary digits (bits)

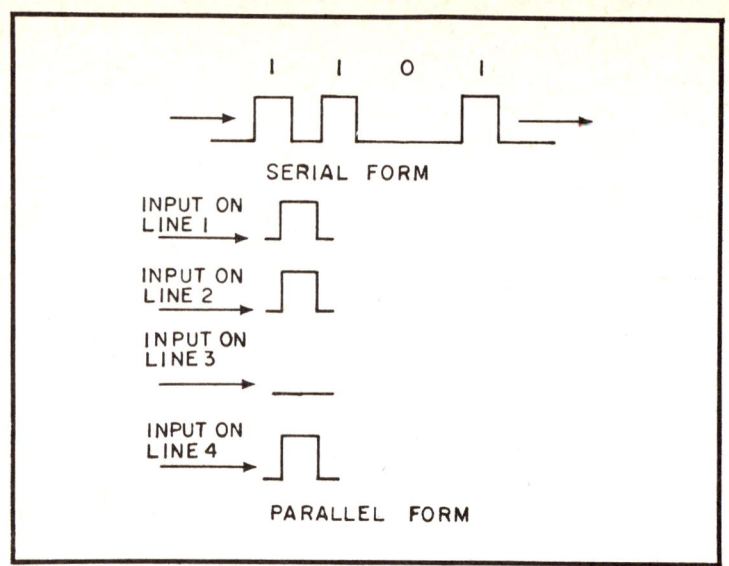

Fig. 5-1. Serial- and parallel-coded pulses.

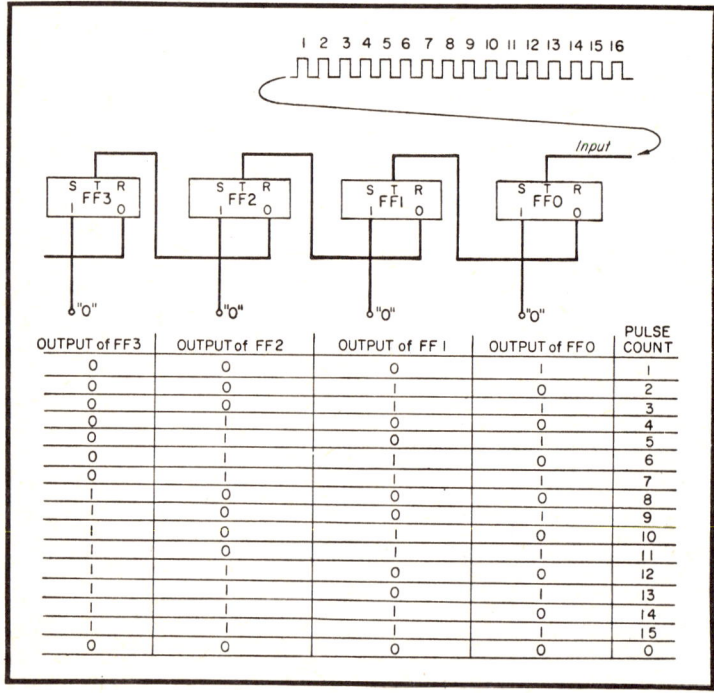

Fig. 5-2. Flip-flop counter.

are received in parallel (one at the set input of each flip-flop). This input is the *augend*. Assume that the number stored in the counter (the augend) is 0011 (3), as shown in the first line below the set terminals, and that the input pulses applied to the toggle inputs are 1010 (10). The counter will add these pulses. In the following explanation of the addition process, it is assumed that the *set* position of the flip-flop, as well as a significant pulse, is represented by a 1 condition. Conversely, a *reset* output, as well as a non-significant pulse input, represents 0.

A delay line is connected between successive flip-flops. The output of each flip-flop is used to provide a significant pulse to the delay circuit on its left only when the set output of that flip-flop changes from 1 to 0 and the reset output changes from 0 to 1.

A 1 input, whether at the toggle input or from the associated delay line, causes a flip-flop to change its state. Thus the flip-flop will change whenever it receives a 1 pulse from either of these sources.

Because the augend is 0011, the following changes will occur when the addend (1010) is received.

    1—FF_2 and FF_0 receive no significant pulse from the toggle input and remain unchanged for the present.

    2—FF_3 and FF_1 do receive 1-state toggle pulses, and both change states. FF_3 now reads 1, and FF_1 now reads 0.

    3—Of the four flip-flops, only FF_1 has changed from 1 to 0. Therefore, it is the only flip-flop to send a 1 pulse (carry) through its delay line to FF_2.

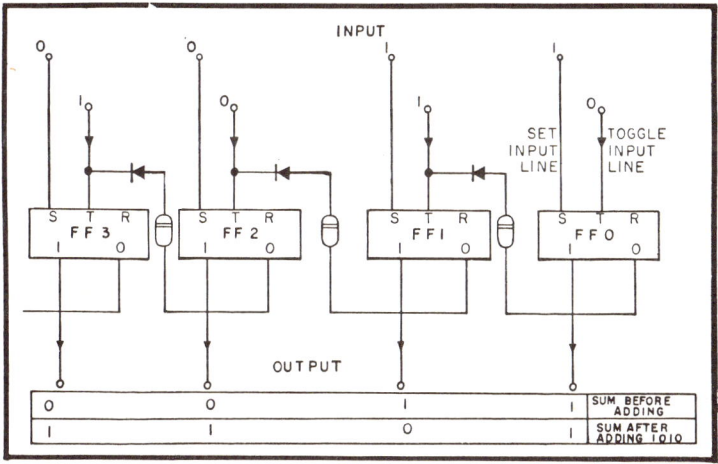

Fig. 5-3. Parallel adder.

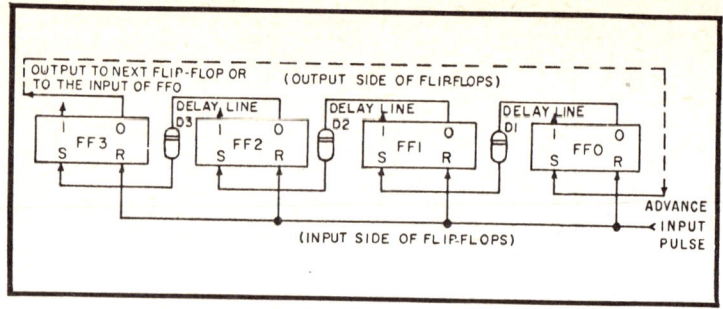

Fig. 5-4. Ring counter.

4—This carry pulse changes FF_2 from 0 to 1. The adder now reads 1101 (13), which is the correct sum of 0011 (3) and 1010 (10).

Note that the diodes prevent pulses from the toggle input (assumed to be positive-going in this case) from interfering with the adjoining flip-flops. The delay lines prevent the input pulses to the adder from appearing at any flip-flop input simultaneously with a carry pulse. In a sense, then, the delay units are used to perform the carry operation.

RING COUNTER

Several flip-flops can be connected to form a ring counter (Fig. 5-4). The name is derived from the fact that the output of the last flip-flop (FF_3) is sometimes connected back to the input of the first (FF_0). This is not a requirement, however, as other means can be provided to initiate the action in the first circuit at the proper time.

In many applications of this circuit, only one of the flip-flops is in the *on* (*set*) condition at a given time. An advance input pulse, which is applied to all flip-flops simultaneously, causes the *on* flip-flop to change its conducting state and, in turn, transmit an input pulse to the next flip-flop to the left. The pulse output from the affected flip-flop is delayed until after the trailing edge of the advance input pulse has dropped to zero, to prevent double-triggering. However, after the delay period, the input pulse to the flip-flop changes the conducting state in that circuit from *off* to *on*.

The next advance pulse will cause the *on* condition to be established in the next flip-flop. This action continues until the *on* condition has advanced from FF_0 to FF_3. A subsequent advance pulse will cause the *on* condition to be transferred either to FF_0 or to the next flip-flop in the ring.

The circuit in Fig. 5-5 is essentially a ring counter to which input paths are added for the purpose of reading in (in parallel) the desired binary digits. As stated before, the four flip-flops make up a register. The action within the ring counter is to shift one digit one place either left or right, but left in this case. An advance input pulse applied simultaneously to each flip-flop in the register will cause the bit previously stored in a stage to be shifted to the next higher-order flip-flop. The bit previously stored in FF_0 will be shifted to FF_1; that previously stored in FF_1 will be shifted to FF_2; the FF_2 bit will move to FF_3; and finally the bit previously in FF_3 will be transferred to FF_0.

If 1011 has been stored before the arrival of the advance pulse, the advance pulse flips FF_3, FF_1, and FF_0 from S to R. Slightly later a pulse from delay line 1 flips FF_1 from R to S. Delay line 2 transmits a pulse that flips FF_2 from R to S. Delay line 3 does not transmit a pulse, and FF_3 remains in the reset condition. Delay line 4 transmits a pulse that flips FF_0 from R to S. Thus, the 0111 condition is stored.

This type of register is called a *shift register*. Left-shift registers are used extensively in computers to perform multiplication. Right-shift registers are used in the division process. Other ways in which flip-flops can be used to perform logic operations are too numerous to consider here. You should therefore study flip-flops and the application of these circuits as presented in this discussion until you are certain of their operation. Only with a thorough understanding will you be able to comprehend other applications of this circuit which are not so elaborately explained.

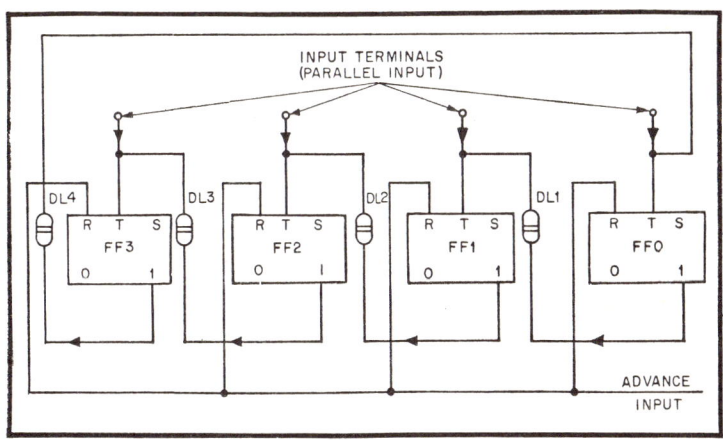

Fig. 5-5. Left-shift register.

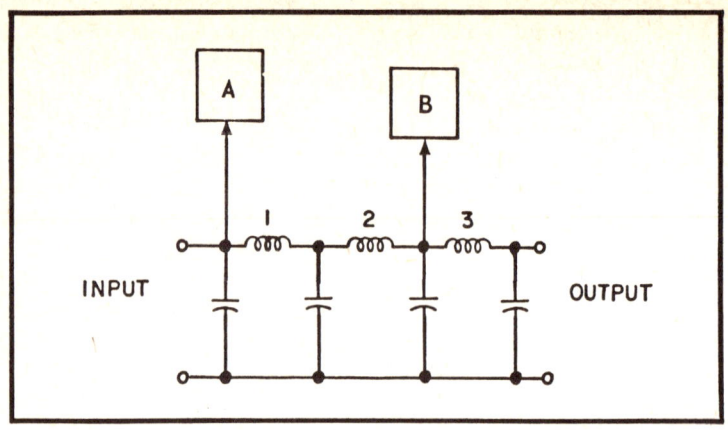

Fig. 5-6. Delay line.

PARALLEL TO SERIAL CONVERSION

The circuit in Fig. 5-6 is called an *artificial transmission line*, or *delay line*. If a voltage is applied to the input terminals of the line, a definite amount of time passes (dependent upon the number of LC sections) before the voltage appears at the output terminals. The LC sections thus give the line the ability to delay the output voltage.

Assume that a voltage must be applied to the circuit in block *B* one or two microseconds after it has been applied to block *A*. This condition can be satisfied merely by constructing sections 1 and 2 of the line for the desired delay.

To avoid the bulkiness of an actual transmission line, an artificial line may be built of coils and capacitors. Such lines have approximately the same characteristics as actual lines but occupy a smaller space. This is the usual method of constructing delay lines.

Now suppose it is necessary to read into a register containing FF_2, FF_1 and FF_0 (Fig. 5-7) in parallel form and to read the information out (at 1.5 μsec intervals) in serial form. One method of performing this operation is described below.

Initially all flip-flops are cleared or reset by an input pulse on the reset line. The information to be stored (in the form of 1s and 0s) is fed over parallel lines to the appropriate flip-flops. A 1 input to any flip-flop will produce a 1 at the *set* output. In the following discussion the *set* output is taken to represent the 1 state of the flip-flop.

Assume that the multivibrator feeds a single pulse to delay line D1 and to AND circuit A1. Because the *set* output of FF_2 is in the 1 state, the coincidence of the inputs to AND element A1 causes this circuit to produce a 1 output during the period t_0. This pulse is read out at J1.

The pulse applied to D1 at time t_0 emerges from the delay line at t_1 and is applied simultaneously to D2 and A2. The presence of this pulse at one input of A2 will not cause the AND element to produce an output (the set output of FF_1 to A2 is 0) and the readout at J1 at time t_1 is 0. The pulse delivered from D2 at time t_2 is applied simultaneously to D3 and A3. Because the *set* output of FF_0 is in the 1 state, AND circuit A3 will produce an output at J1 at time t_2. Thus the special readout of the register is accomplished, at the specified intervals, as 101.

TIMING CIRCUITS

Timing circuits in the computer insure proper timing of events. Timing pulses *enable* some circuits, and thereby permit certain operations to begin and certain other operations to be terminated. A few microseconds later the return of the enabling pulses to some new state signifies the expiration of time for the selected operations. Immediately after disabling one group of circuits, enabling timing pulses are applied to others, and another set of operations is performed. This process of enabling

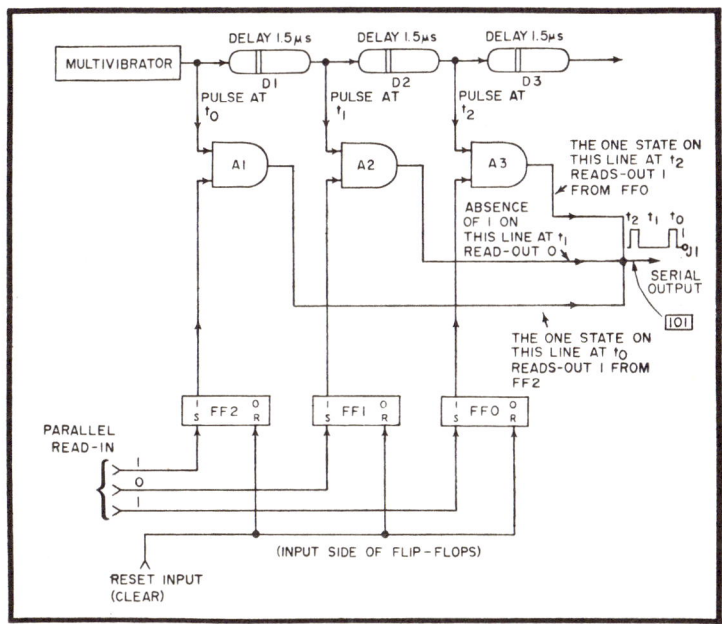

Fig. 5-7. A method of using delay lines for serial readout.

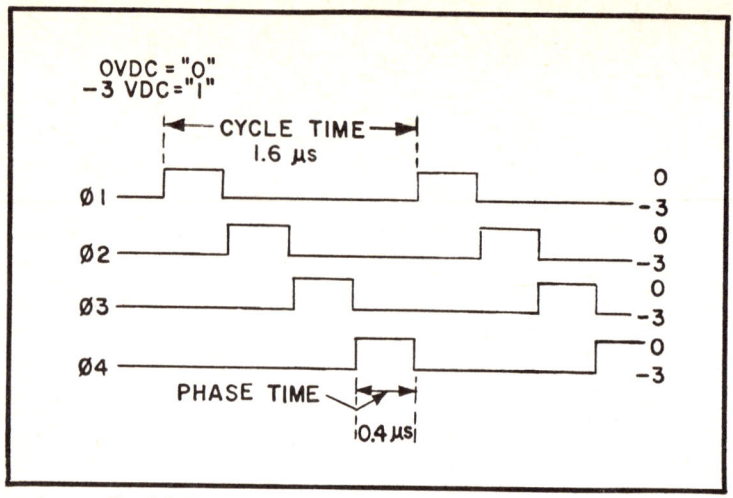

Fig. 5-8. Timing sequence for 4-phase master clock.

some circuits and disabling others in sequence is repeated over and over again until the program has been completely executed.

CLOCK PHASES

A circuit arrangement called a *master clock* provides the main timing signals within a digital system. Some of the characteristics of this circuit are treated below.

The master clock system (Fig. 5-8) utilizes a 4-phase output with a complete cycle time of 1.6 μsec. The four clock phases are actually the master clock output pulses, arranged in groups of four. The four phases are obtained from an arrangement of gates and multivibrators in such a way that only one phase is high in any given 0.4 μsec period.

ENCODERS AND DECODERS

It has been stated that digital circuits are designed to operate on data that is in the binary form; that is, either 1 or 0. In most computers it is not practicable to enter numbers and other information in binary form, as this would require the programer to spend too much time in the detailed effort of accurately representing large numbers of complicated alphabets and symbols. Likewise, it is not usually desirable to present the final output of a digital circuit in binary form, as this requires too much time in reading and interpreting. Thus it is necessary to perform conversion on both input and output information.

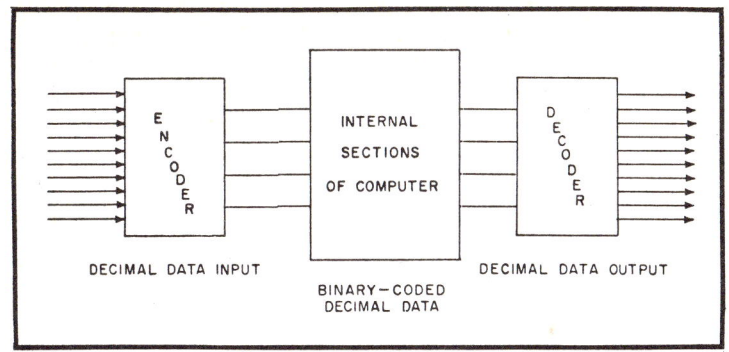

Fig. 5-9. Basic principle of encoding and decoding.

Digital equipment is generally equipped with encoders which change discrete inputs into a combination of coded outputs, and decoders which transform the internal binary data to its more conventional form at the output. Thus, the programer can use the familiar decimal system or the octal system to program a computer, depending on the encoder used to convert the data to the binary form as required internally by the computer. The output of the internal circuits of the computer is then changed by the decoder back into the familiar decimal form. The basic principle involved is illustrated in Fig. 5-9.

The simple decimal-to-binary encoder (Fig. 5-10) receives any one of 10 decimal numbers at its input and produces a binary-coded decimal output. This action can be seen by a study

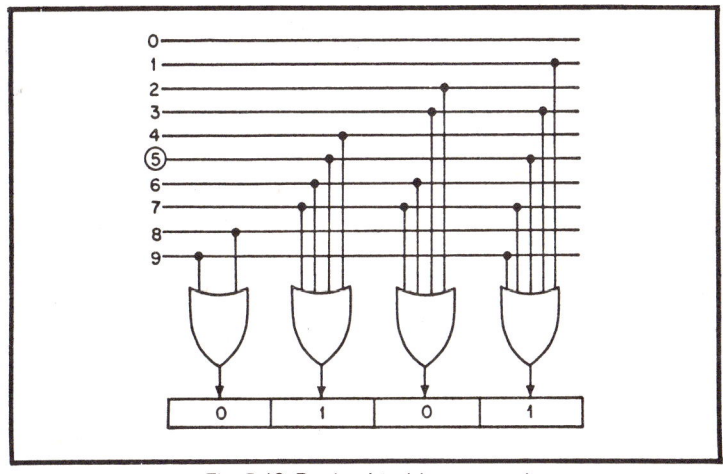

Fig. 5-10. Decimal-to-binary encoder.

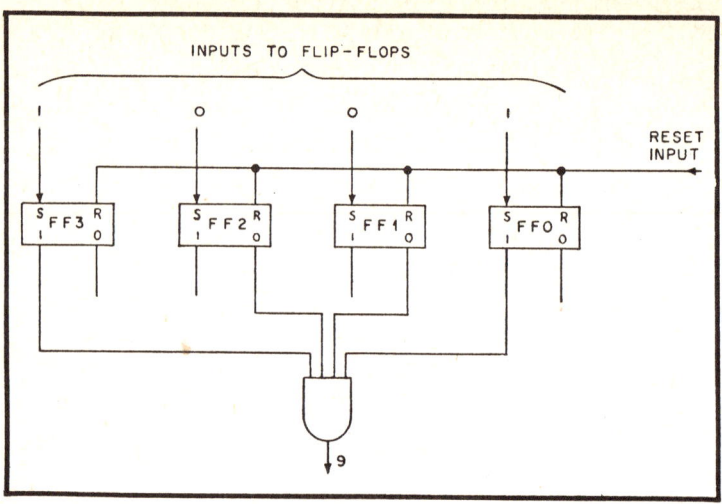

Fig. 5-11. Binary-to-decimal decoder.

of the circuit using OR logic. (There are more complex types of encoders.)

The basic principle of binary-to-decimal decoding (Fig. 5-11) uses flip-flops to provide the inputs to the AND element, which, in turn, produces the decimal output. This type circuit produces an output only when the right combination of flip-flop outputs is applied to the inputs of the AND element. Appropriate use of either inverters or flip-flop output connections makes possible the decoding of any binary-coded decimal digit. Although only binary-coded decimal conversions are illustrated here, similar procedures can be used to encode and decode numbers in all number systems.

A more advanced form of encoding and decoding uses a *matrix* to translate between number systems. A matrix is an array of circuit components such as wires, diodes, magnetic cores, and relays. The matrix performs a specific function, such as converting from one numerical system to another. In mathematics, *matrix* means a set of terms which operate on one type of number to produce a second type—in other words, a number translator.

As discussed earlier, binary arithmetic uses only the two digits, 1 and 0. The manner in which these digits are arranged determines their numerical value. The advantage in using the binary system in digital equipment is that the two digits (1 and 0) can be easily represented by the two possible stable states of certain electronic circuits and components.

The operation of a matrix is based upon certain principles which were first investigated by mathematician James Sylvester in the middle of the nineteenth century. You will recall that for any particular circuit previously discussed, a single output is produced if some specific combination of inputs exists. Further, there is a fixed number of possible input combinations (depending upon the number of inputs used) each one of which is called a *minterm*.

A circuit with two inputs, for example, has four minterms: AB, $\overline{A}B$, $A\overline{B}$, and $\overline{A}\overline{B}$. A 3-input circuit has eight minterms: ABC, $AB\overline{C}$, $A\overline{B}C$, ABC, $A\overline{B}\overline{C}$, $\overline{A}\overline{B}C$, $\overline{A}B\overline{C}$, and $\overline{A}\overline{B}\overline{C}$. Thus, it can be said that for n inputs, there are 2^n minterms. Consequently, for 2 inputs there $2^2 = 4$ minterms; for 3 inputs, there are $2^3 = 8$ minterms; for 4 inputs, there are $2^4 = 16$ minterms; etc. Each minterm indicates one of the possible input combinations that can occur.

The logic matrix differs in one respect from other circuits discussed in this chapter in that a matrix with n inputs has 2^n outputs—one output for each minterm. It produces a signal at one specific output terminal for any one combination of inputs.

Consider the simplified illustration in Fig. 5-12. The matrix has three inputs (A, B, and C), each of which can be in either the 1 or 0 state. Each combination of digits represents a binary number, so that ABC is interpreted as 111, $\overline{A}\overline{B}\overline{C}$ means 000, $\overline{A}B\overline{C}$ means 010, etc. In the illustration, a signal exists at A and at B, but not at C, representing 110. The appearance of this

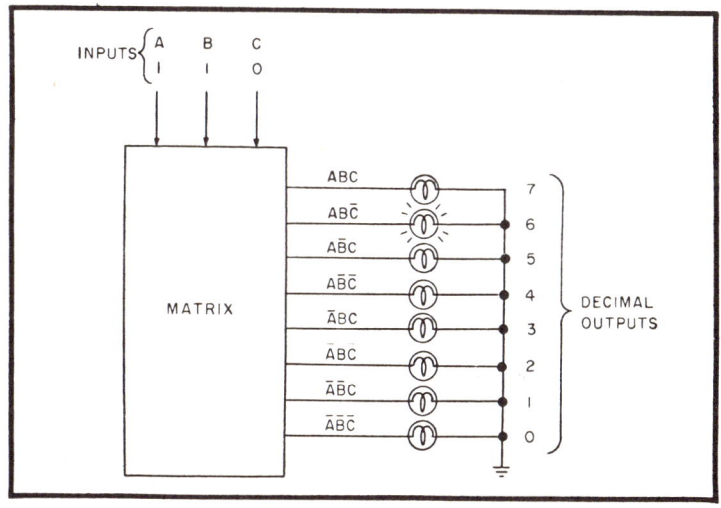

Fig. 5-12. Three-input matrix.

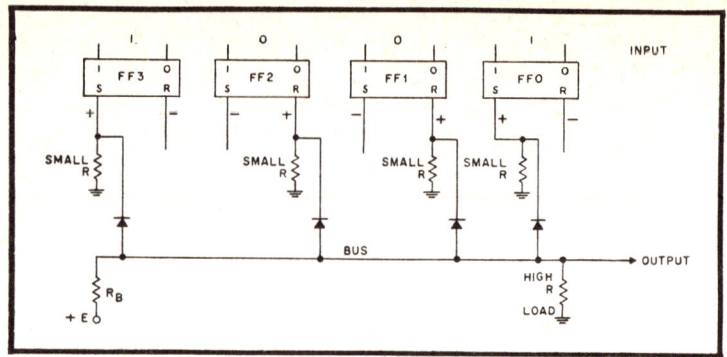

Fig. 5-13. Simple flip-flop decoder.

combination of signals at the input will produce a signal on one specific matrix output. By connecting this output to a numbered indicating device, such as a lamp, the binary number 110 is directly converted to its decimal equivalent, 6.

In general, each minterm represents a specific number. For example, $ABC = 000 = 0$, $ABC = 001 = 1$, $ABC = 010 = 2$, and so on to $ABC = 111 = 7$. To decode higher numbers, the number of inputs and outputs must be increased. Because a logic matrix has an output for each minterm and every minterm represents a binary number, each output can be labeled with the appropriate decimal value.

Consider the circuit in Fig. 5-13, which is used to decode the binary-coded decimal number 1001 (9). When the magnitude of the voltage at any one of the flip-flop outputs decreases below the bus voltage, one or more of the diodes conducts. Because the resistances are small and in parallel, most of the applied voltage is distributed across R_B. When, and only when, the flip-flops are in the condition shown—that is, from left-to-right: set, reset, reset, set—all diodes will be cut off. And because of the high R load, the voltage at the output will rise in the positive direction. This rise in voltage indicates that a binary-coded decimal 9 is stored in the flip-flops. Similarly, any binary number can be decoded by connecting the diodes to the proper flip-flop terminals.

A more versatile form of diode matrix is illustrated in Fig. 5-14. Here, diodes are connected to flip-flop terminals in a manner that will decode any one of eight conditions representing 000 through 111. The diodes are normally conducting (low voltage or zero output) except when the associated flip-flop output is positive (which reverse biases the diode).

132

A 0 (rise in voltage) is produced at the matrix output when all flip-flops are in the *reset* condition. A 1 output is produced when both FF_2 and FF_1 are reset and FF_0 is set. A 5 is produced when FF_2 and FF_0 are set and FF_1 is reset. In other words the *set* outputs of FF_2 and FF_0 and the *reset* output of FF_1 are positive, so that all diodes on the 5-level are cut off and the output voltage on that level is positive.

A study of the matrix for various combinations of flip-flop outputs reveals the conditions that exist when representing any decimal digit from 0 to 7. Higher order numbers can be decoded by increasing the number of flip-flops and connecting additional diodes to indicate higher counts.

SYNCHRONOUS CONTROL

Every computer seems to have some unique control feature peculiar to that particular machine. There is no one so-called "best method," since the control method cannot, in most cases, be decided without considering several other factors, such as storage-access time, input/output devices, and the nature and time required to perform arithmetic operations.

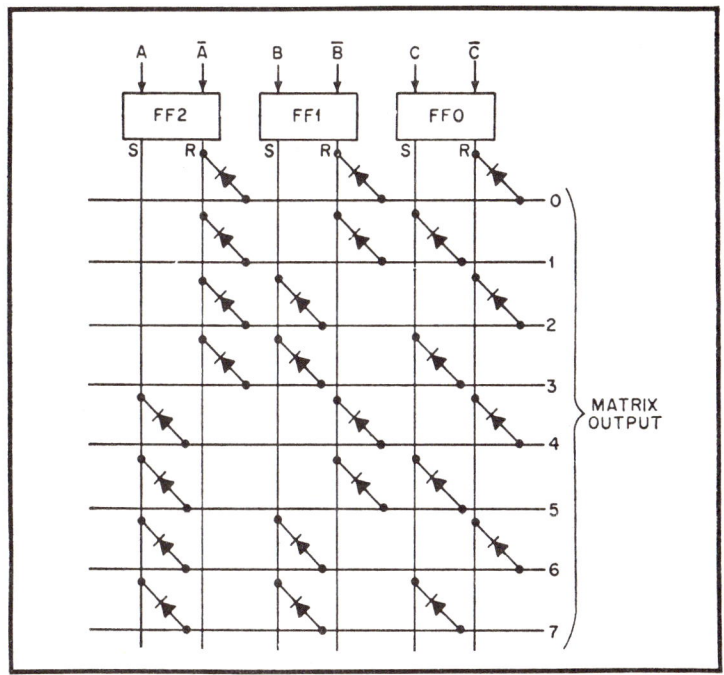

Fig. 5-14. Diode decoder matrix.

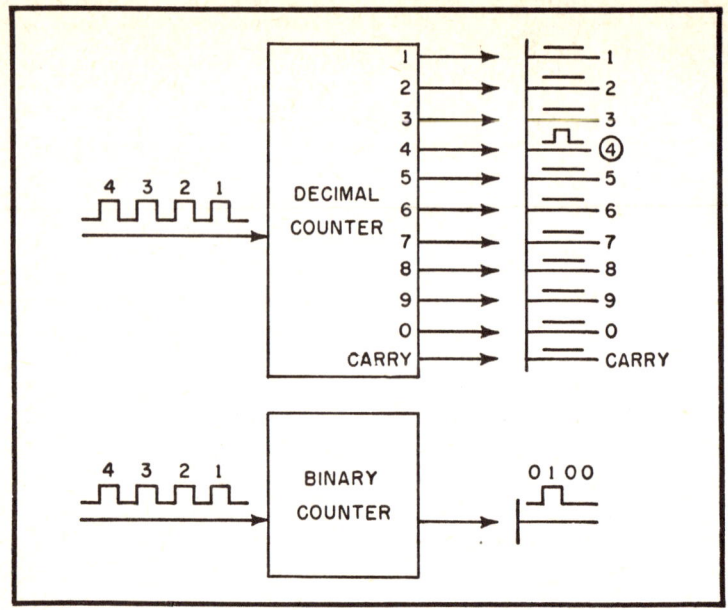

Fig. 5-15. Unitary code conversion to decimal or binary code.

Synchronous control is a mode of computer operation characterized by a fixed time period for the execution of each operation. Conversely, *asynchronous* control uses varying amounts of time to execute its operations, depending of course on the type of operation being performed. In the asynchronous control method, the advance to the next command is signaled when the execution of the previous command has been completed.

UNITARY OPERATIONS

Unitary operations make use of the *unitary code*. The unitary code has only one digit, and the number of times that this digit is used in a number determines the quantity it represents. Thus

$$1 = 1$$
$$11 = 2$$
$$111 = 3$$
$$1111 = 4$$

It can be seen therefore that this is not a very practical code for representing a large number such as 10,000. The unitary code is mentioned here only because it is required for the explanation of counters that follow.

COUNTERS

The unitary code is converted to a coded system of numerical notation when applied to the input of counters. For example, in Fig. 5-15, the unitary number 4 is converted to the decimal 4 (indicated by some readout device, like a lamp or odometer). When the unitary number 4 is applied to a binary counter, the number 4 is counted in pure binary (8-4-2-1 code). There are many different types of counters, some of which are discussed below.

The capacity of a counter to count to a certain numerical value is known as the counter's *modulus*. Thus, we speak of *modulo-5* and *modulo-10* counters in defining the maximum quantity that the counter in question is able to count to.

Let us examine the odometer, the mileage indicator in an automobile. Most automobiles have odometers that have a decimal readout capacity of 99,999.9 miles. But, the modulus of this odometer is 100,000.0. Since 99,999.9 is the maximum readout, the recycling of the readout indicator to 00000.0 represents 100,000.0. Thus the modulus of any counter is the capacity of the counter plus one in the least significant digit.

Another example of a decimal readout counter is a people counter. This hand counter is used in counting the number of people that pass through a place at a given period. The pressing of a button on the counter advances the counter by one. Its modulus would also be the fixed capacity of the counter plus one.

The binary number system is used in all digital systems. Thus, the type of counter used most frequently in digital systems is the binary counter.

Binary Up Counters

A binary counter is merely a series of flip-flops. The number of flip-flops used determines the modulus of the counter. Figure 5-16 shows a 4-stage binary counter. The input pulses (in unitary code) are applied to the flip-flop corresponding to the lowest-order binary number. Note that the C (clear) input from the 0-side of each flip-flop feeds the input of the next (higher order) flip-flop. The S (*set*) from the 1-side of each flip-flop is in the *set* position, the output from the 1-side of the circuit reads 1; if the circuit is in the *clear* position, the 1-side output reads 0. Every time the *clear* side changes from a 0 to 1 state (1 to a 0 on the *set* side), a pulse is sent to the next higher-order flip-flop input. An input to the counter shown in Fig. 5-16 affects the flip-flops in the following manner:

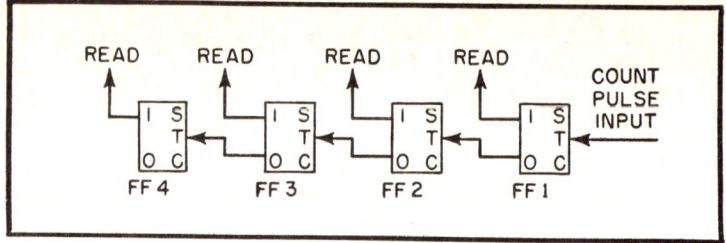

Fig. 5-16. Four-stage binary counter.

1—Assume that all four flip-flops are in the *clear* position. Each output reads 0 and the total counter output is 0000 (binary 0).

2—A single pulse is introduced. FF_1 changes to its *set* position, producing a 1 at its output. However, the signal is not coupled to FF_2 because the flip-flop produces a pulse at its *clear* output only when changing from *set* to *clear*. The counter's output now reads 0001 (binary 1).

3—Another pulse enters the counter's input. FF_1 changes from *set* to *clear*, producing a 0 at its *set* output. This time, a pulse is produced at the *clear* output (*clear* side changes from 0 to 1). The *clear* output pulse is coupled to FF_2, causing it to change from the *clear* to the *set* position. The FF_2 *set* output now reads 1. (Again, because each flip-flop couples a pulse only when going from the *set* to *clear* position, no pulse is produced from the FF_2 *clear* output, and the remaining flip-flops are not affected.) The counter's output now reads 0010 (binary 2).

Each successive input pulse to the counter affects the flip-flops in a similar manner. The third input pulse will change FF_1 to the *set* state, producing 0011 (binary 3). The fourth pulse will change FF_1 and FF_2 to the *clear* position and FF_3 to the *set* position, producing 0100 (binary 4). Continuing this analysis reveals that the circuit will count up to 15 pulses, indicating its tally in binary code.

Thus, the highest possible binary number in the counter's output is 1111 (binary 15). The modulus, therefore, is $15 + 1 = 16$. Since FF_4 counts to 16, the addition of another flip-flop will permit counting to 32.

Binary Down Counter

To perform certain functions, digital equipment may be required to cycle through a predetermined number of operations and then stop. A circuit used to count down from a preset number is the *decrementing* or *down counter*. Such a circuit is illustrated in Fig. 5-17. This circuit is a module-16 down counter. It will count down from 1111 to 0000. The operation of this circuit may be understood by considering all the flip-flops to be set to the *one* state (1111) and referring to the waveforms depicted in Fig. 5-18.

Symbolized Ring Counter

Another type of counter is the ring counter. One type of ring counter was treated earlier. The ring counter is cyclic, having no beginning and no end. After it has reached its highest number, its next count is 0. In the ring counter, only one element or state is on at one time. Each input signal advances the *on*, or operating, stage one step forward.

The ring counter shown in Fig. 5-19 uses flip-flops and AND circuits. Note that the input pulses never enter any flip-flop directly. Instead they are applied to all of the AND circuits. An input pulse is used by each AND circuit to clear the flip-flop corresponding to that AND circuit (if the flip-flop is in the *on* state), and to set the following flip-flop.

If we assume that the *on* signal is currently in flip-slop FF_1, the operation would be as follows: When an input pulse is applied, AND circuit A1 is the only one that produces an output signal because A1 is the only AND circuit that has a *set* output signal at its second input. Note that this output signal branches and affects both FF_1 and FF_2. The output signal of AND circuit A1 clears FF_1 to 0. This flip-flop is therefore no longer *on*. The output signal from A1 also sets FF_2 to its 1 position; FF_2 is now on. AND circuit A2 is now ready for the next input pulse from the input bus, since it already has a 1 at its lower terminal.

Note that the connections between flip-flops FF_2 and FF_3 are identical to the connections between FF_1 through the last flip-flop and back to FF_1.

Thus it can be seen that the counter uses a cyclic principle, and that only one stage is *on* at any one count. The modulus of a

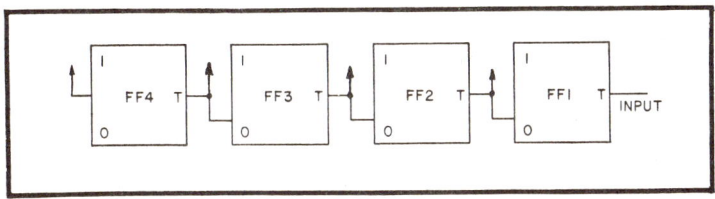

Fig. 5-17. Down counter.

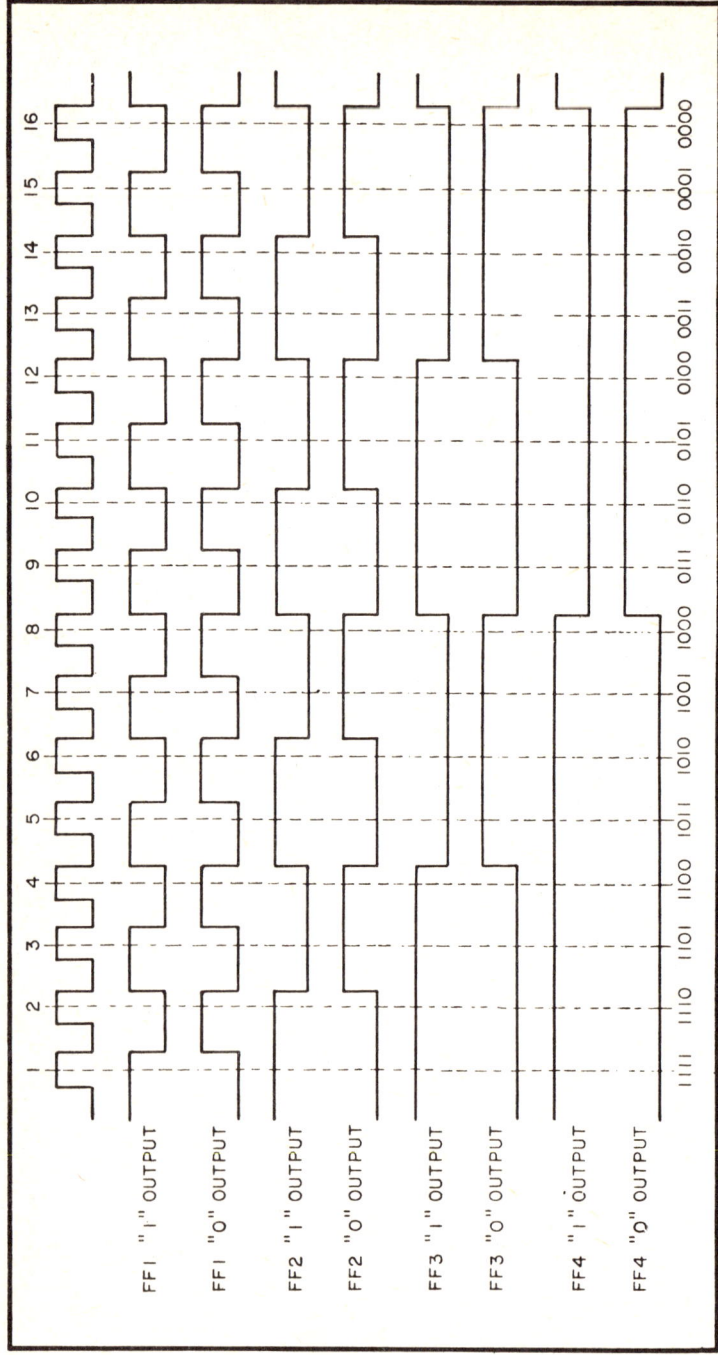

Fig. 5-18. Timing diagram of Fig. 5-17 circuit.

ring counter can be increased by adding more elements (flip-flops and AND circuits in this case).

FREQUENCY DIVISION

Binary counters are also used to count the number of input pulses to the circuit and produce an output after a certain number of pulses are received. In this way the counter divides the input frequency by some factor and is therefore referred to as a frequency divider. The function of a binary frequency divider or counter is to divide, or scale down, the number of input pulses by some power of 2. A single flip-flop, for example, divides by 2. It produces one output pulse for every two input pulses.

The output pulses can be either *in phase* with or *out of phase* with the count pulses. The input and output in Fig. 5-20 are positive. Flip-flop FF_1 can be made to produce a positive output so that a second flip-flop of the same type can, upon receiving the output of the first, again divide the input by 2. Thus the output of a 2-stage circuit produces only one pulse for every four pulses at the input, making it divide-by-4 circuit. Similarly, a third flip-flop will make it a scale-of-8 divider, producing one output pulse for every eight input pulses.

Figure 5-21 shows a 4-stage binary scaler and the pulse outputs of each stage. Note that each flip-flop produces only half as many pulses as the one before it, so that the final stage (FF_8) produces only 1 pulse for every 16 pulses introduced at the first stage. The circuit is therefore a scale-of-16 divider.

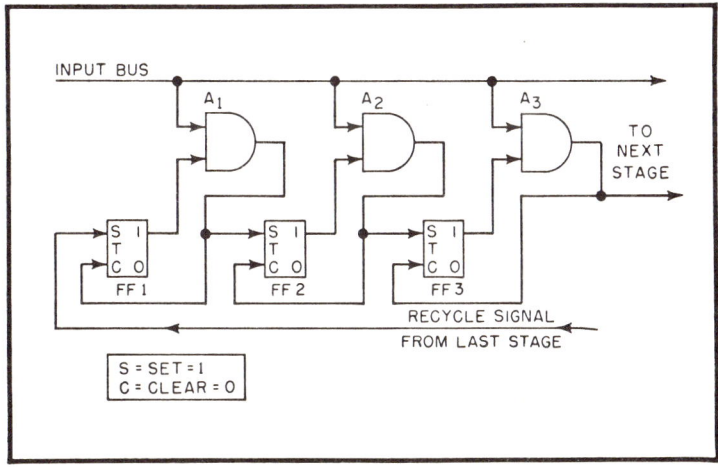

Fig. 5-19. Ring counter using flip-flops and AND circuits.

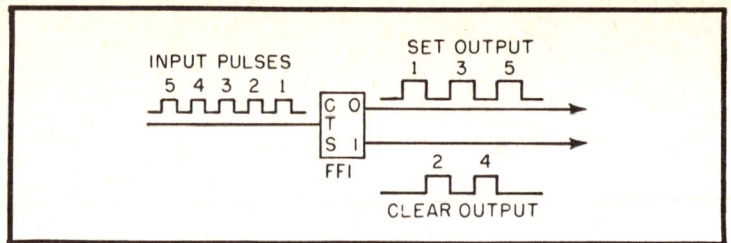

Fig. 5-20. Using a flip-flop as a scale-of-2 divider.

INFORMATION REPRESENTATION AND TRANSFER

The advantage of parallel operation is that data is transferred more quickly than in serial operation. However, in serial operation, fewer circuits are required to perform the same operation. The choice between the two methods is one of design. Although any storage device can be adapted to either method, some types are best suited to one method or the other.

Serial and parallel techniques are often used in the same system during different operations. For example, a data word may be transferred to a *shift register* in parallel and then read out of the register serially. The reverse is also true; a data word may be read into a register serially and read out in parallel. The

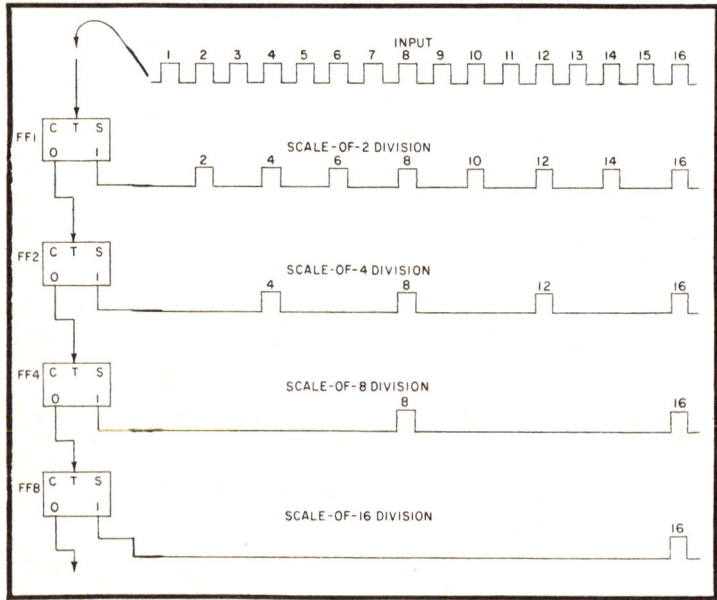

Fig. 5-21. Pulse train of a 4-stage binary scaler.

following is an explanation of the operation of a shift register using serial and parallel techniques.

SHIFT REGISTERS

As explained in earlier sections, a register is a temporary storage device. Data is put into a register and can be read out of the register when needed for some operation. Shift registers can shift the data bits to the right or to the left, operations that can round off a number, or multiply or divide by 2, 4, 8, 16, etc. The shift-left and shift-right operations may use serial and parallel techniques. Figure 5-22 is a shift register that illustrates the concepts of right and left shifting using parallel techniques.

Right Shift

Assume that information is stored in the A-register (Fig. 5-22). It is now desired to shift the information in the A-register (represented by the 1 state on the *set* sides of the flip-flops) to the right by one place. The clear pulse is applied to the clear sides of the X-register flip-flops to clear the register of any information present from a previous operation. Note that the clear pulse is applied in parallel to all the flip-flops in the X-register.

If a shift right pulse is applied in parallel to AND circuits A2, A4, A6, and A8, the data from the *set* sides of the A-register flip-flops is transferred to the X-register, and, in the transfer process, is shifted one place to the right. This operation uses both a parallel and serial movement of data. All the information in the A-register is transferred to the X-register in parallel, while the shift motion is serial.

For example, the output of the *set* side of FF_{A3} is ANDed with the shift right pulse at AND circuit A2. The output of this circuit is passed by OR circuit O2 to the input of the *set* side of FF_{X2}, thereby effecting transfer of a data bit from the first flip-flop in the A-register (FF_{A3}) to the second flip-flop in the X-register (FF_{X2}). Similarly, data from FF_{A2} is transferred to FF_{X1} and from FF_{A1} to FF_{X0}. The LSD (least significant digit) from FF_{A0} is lost during a right shift operation.

Next, the clear pulse is applied to the clear sides of the A-register flip-flops, thereby clearing these flip-flops of that data originally stored in the register. The transfer pulse is then used to transfer the shifted data, now in the X-register, to the corresponding flip-flops in the A-register by way of AND circuits A9, A10, A11, and A12. The transfer pulse is ANDed with the

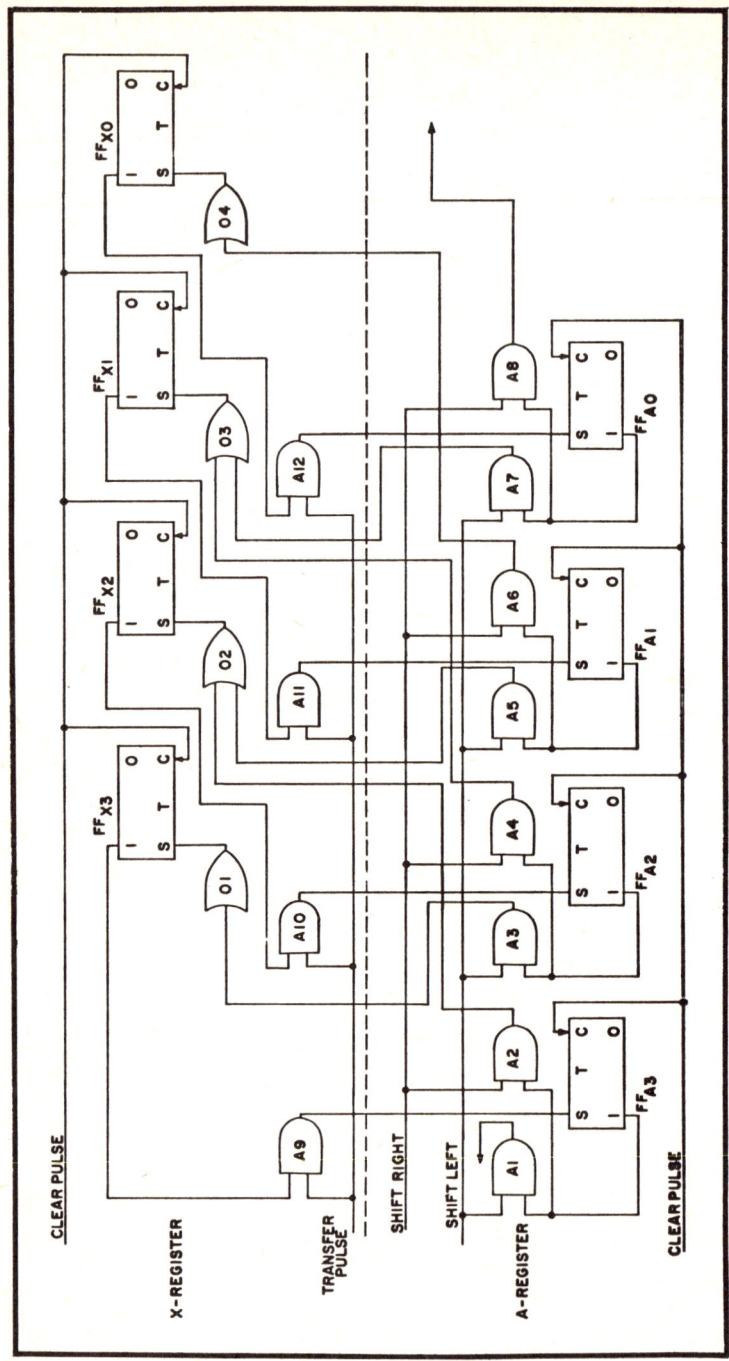

Fig. 5-22. Example of shift register used in arithmetic section of computer.

outputs of the *set* sides of the X-register flip-flops. The outputs of these AND circuits are applied to the corresponding *set* sides of the A-register flip-flops. Thus data are now stored in FF_{A2}, FF_{A1}, and FF_{A0}, and a shift, one place to the right, has occurred. Note that FF_{A3} is not set during the transfer back to the A-register, since FF_{X3} remained in the cleared state during the original transfer from A to X.

Left Shift

The left-shift operation is similar to the right-shift operation, except that now AND circuits A1, A3, A5, and A7 are used for transferring and shifting. The X-register is cleared by the application of a clear pulse to the *clear* sides of the X-register flip-flops. Then, the shift left pulse is applied in parallel to A1, A3, A5, and A7. The outputs of the *set* side of the A-register flip-flops are passed by the AND circuits (which simultaneously received a left shift pulse), and OR circuits (O1 through O4) to the *set* sides of the X-register flip-flops. Thus, data is transferred and shifted from FF_{A0} to FF_{X1}, from FF_{A1} to FF_{X2}, and from FF_{A2} to FF_{X3}. Usually, during a left shift, the most significant digit is shifted to the least significant digit position; i.e., FF_{A3} to FF_{A0}.

A clear pulse to the A-register flip-flops clears the register. The transfer pulse now enables the contents of the X-register to be transferred to the corresponding A-register flip-flops by way of A9, A10, A11, and A12, respectively. Note that the data is now stored in the A-register, one place to the left of its original position.

BINARY ADDITION

The addition operation is performed in the arithmetic section by the *binary adder*. The axioms of binary addition are

X	Y	f_S	f_C
0	0	0	0
0	1	1	0
1	0	1	0
1	1	0	1

where X and Y are variables, f_S represents the sum, and f_C represents the carry.

Half-Adder

A half-adder (Fig. 5-23) is a circuit which has two input points, identified here as X and Y, representing addend and

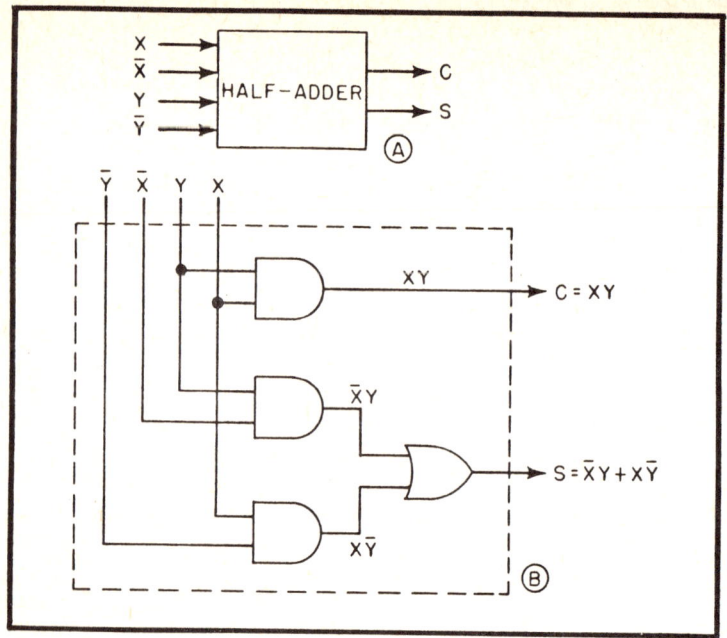

Fig. 5-23. Half-adder.

augend. There are also two output points, S and C, representing sum and carry, respectively. This half-adder performs each of the addition axioms cited above. Each of the variables, X and Y can assume a value of either 0 or 1. If the X and Y inputs are from flip-flops, these inputs and their complements are available. In binary addition, the Boolean expression for the sum equal to 1 is

$$S = X\overline{Y} + \overline{X}Y$$

The Boolean expression for the carry is

$$C = XY$$

Full-Adder

The full-adder circuit in Fig. 5-24 is designed to advance a carry (in serial addition) through as many columns as necessary. This action is accomplished by circulating a carry digit through the carry-feedback circuit comprising A2, CR2, and D1, each time a logic-1 condition exists at the A2 output. The following example will illustrate this feature.

Consider the addition of 111 and 011 as shown. The serial columns are applied at times t_0, t_1, and t_2 respectively. OR circuit

144

G1 produces a 111 input to one terminal of inhibitor I1, with one pulse applied during each time interval. AND circuit A1 produces an inhibitory input to I1 during time intervals t_0 and t_1, so that the output of I1 is 100.

The carry digits (011) from A1 are also fed through CR1 (an isolating diode) to delay line D1 where each digit is delayed one bit time. The output (110 corresponding to time intervals t_2, t_1, and t_0, reading from right to left) is applied to the lower input terminal of G2, causing the output of this circuit to be 110 during the time intervals shown.

Note that the input to G2 from I1 during the intervals t_2, t_1, and t_0 is also applied to one terminal of AND circuit A2, and that the first carry output (110) is applied to the other terminal. Thus A2 produces a 100 output during intervals t_2, t_1, and t_0.

The 1 output of A2 during period t_2 causes the 1 input to I2 from G2 during the t_2 interval to be inhibited, and the output of I2 during this period is 0. The I2 output during intervals t_0 and t_1 appears uninhibited. Thus the I2 output from t_2 to t_0 is 010.

The presence of a 1 in the A2 output indicates that a carry is yet to be added to one of the remaining columns before the true sum can be produced. The carry digit is fed through CR2 to D1 where it is delayed and shifted into the period t_3. The second output of D1 (shown as the second carry input to G2) is passed through G2 during the t_3 interval and combined with the serial train already at the output. This action produces the true sum (1010) of the binary inputs at G1. Note that a carry has been advanced from the first column to the fourth. In a similar manner this circuit can advance the carry through any number of columns as required.

A parallel adder circuit which uses flip-flops as the basic elements is treated elsewhere. In the discussion of parallel addition, it is shown that a carry digit is generated when any flip-flop changes from the *set* to the *reset* condition and that this

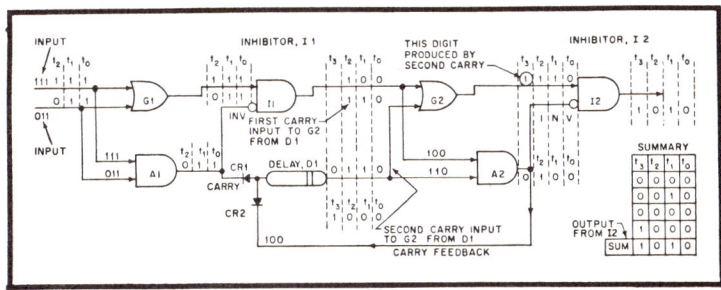

Fig. 5-24. Full-adder for accumulating multiple carry digits.

Table 5-1. Truth Table of f (A, B), Showing Previous Carry, New Carry, and Sum.

A	B	C	S	C_n
0	0	0	0	0
0	0	1	1	0
0	1	0	1	0
0	1	1	0	1
1	0	0	1	0
1	0	1	0	1
1	1	0	0	1
1	1	1	1	1

carry is fed through a delay circuit to the next higher order. The delay circuit prevents the accumulation of the carry digit in the next higher column until after the addition of the augend and addend in that column.

The axioms treated earlier for the addition of two binary numbers do not include all possibilities. When the addend and augend contain more than one column, a circuit designed to execute such additions must be able to accept and add a carry digit from a previous column. Table 5-1 shows all possibilities which may be encountered in the addition of such numbers, where C represents a previous carry which is now to be added to the sum of the addend and augend in that column, C_n represents a new carry which is to be added to the next column, and S represents the sum.

From the truth table (Table 5-1), it can be seen that the minterm equation (condition of the variables for which $f = 1$) for the sum (S) is

$$S = \overline{A}\overline{B}C + \overline{A}B\overline{C} + A\overline{B}\overline{C} + ABC$$

and that the minterm equation for the new carry digit (C_n) is

$$C_n = \overline{A}BC + A\overline{B}C + AB\overline{C} + ABC$$

The truth table (and consequently the equations) indicates that the sum is 1 when only one input (either A, B, or C) is 1, or when all inputs (A, B, and C) are 1. The condition when A and B are 1 is not present in the sum equation since the sum output for

146

this condition is 0. The C_n equation indicates that a new carry is generated whenever any two of the variables are 1, or when all variables (A, B, and C) are 1.

The C_n equation can be simplified to

$$C_n = AB + AC + BC$$

AND or OR circuits can be combined as shown in the lower portion of Fig. 5-25 to produce a 1 condition at the sum output when any one of the conditions shown in the sum equation exists at the input. Likewise, the logic circuit arrangement in the upper portion of the figure produces a 1 carry output to the next higher order when any of the conditions contained in the carry equation exist at the input.

Remember that this circuit is designed to accommodate a single column with only two binary digits plus a carry digit. The

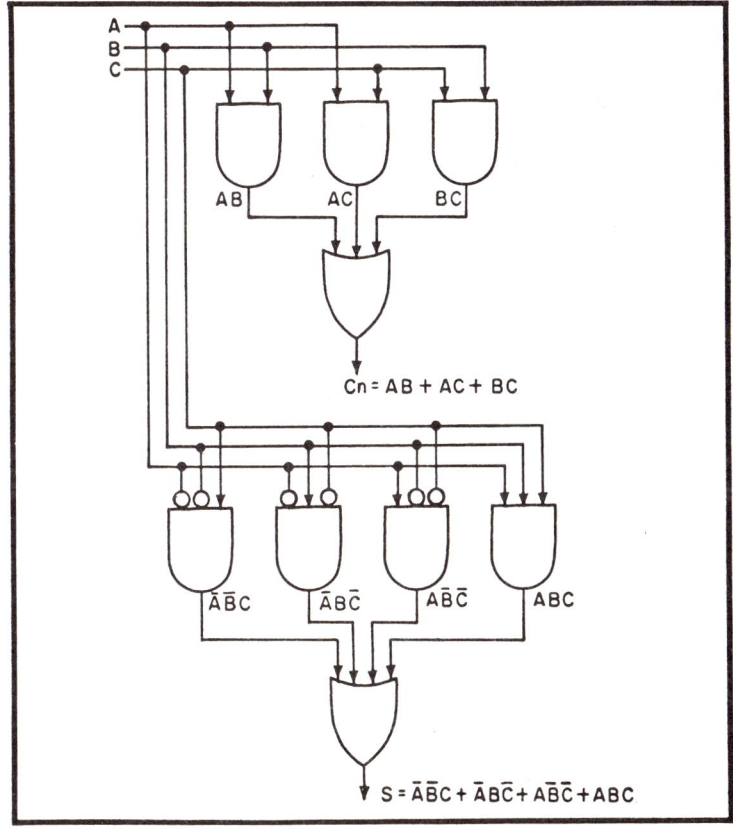

Fig. 5-25. Three-input adder.

147

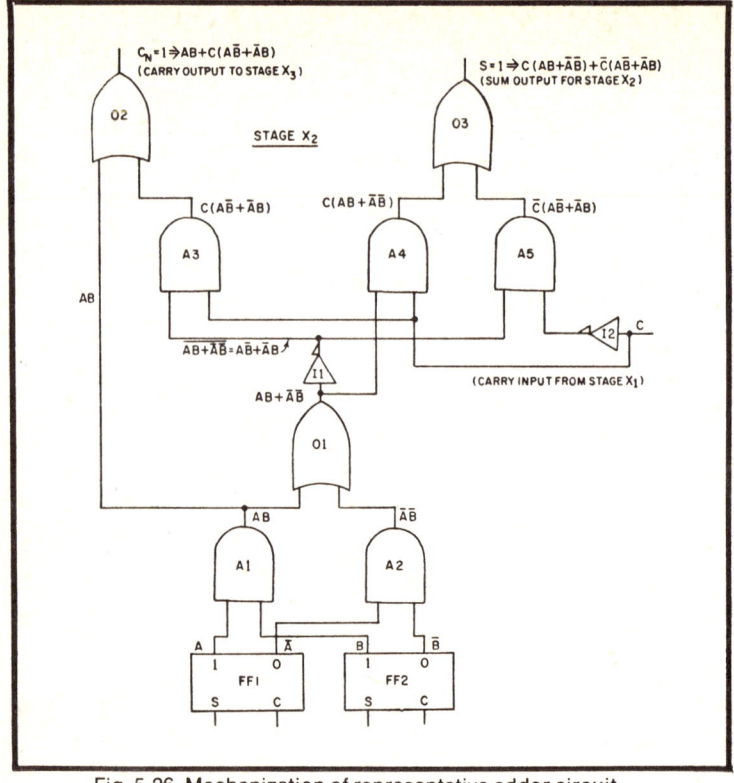

Fig. 5-26. Mechanization of representative adder circuit.

addition of several columns of digits would require a similar circuit for each column (for parallel operation) or sufficient time between the application of binary inputs during serial operations to permit the carry function to be executed.

Full-Adder Using AND/OR Logic

The equation can also be mechanized as shown in Fig. 5-26; this method is more commonly used in computers. To analyze the operation of this circuit, consider the output of the flip-flops FF_1 and FF_2 for a condition when A and B are high and when carry input C is low. In this case, it is expected that the sum output of the stage X2 (the stage under discussion) will be low, and that the carry output will be high. These outputs can be verified by noting that the high output for A of FF_1 and the high output for B at FF_2 enable A1 to produce a high output. This high output passes through OR circuit O2 and is applied as the carry input to the next higher stage (stage X3 not shown).

148

The A1 output is also applied through O1 to AND circuit A4 as a *partial enable* for this circuit. Because the carry input from stage X1 (the next lower order stage, not shown) is assumed to be low, A4 is not enabled and a low signal is produced as output. The low-level C is inverted by I2 to a high level, which partially enables A5. However, the high-level $AB + \overline{A}\overline{B}$ signal at the O1 output is inverted by I1 to a low level, and A5 is not enabled. As a result, both inputs to O3 are low, and the O3 output is low.

Now assume that the A, B, and C inputs to stage X2 are high. The carry is produced in the same manner just described. Because C is also high in this case, A4 is enabled and the function C $(AB + \overline{A}\overline{B})$ is high. Thus the high output of A4 passes through O3 to represent a 1 condition sum for the stage X2 output.

Consider a situation where A = high, B = low, and C = high. The outputs of A1 and A2 are low, resulting in a low output from O1, which disables A4. The low input to inverter I1 is inverted and the resulting high output, representing an active complement of $AB + \overline{A}\overline{B}$ $(A\overline{B} + \overline{A}B)$ is applied as a partial enable to A3 and A5. Because C is high, the I2 output is low and A5 does not become fully enabled. However, the C input (high in this case) is applied directly to A3, so that with the high input from I1, A3 is enabled. In this condition, the function $C(A\overline{B} + \overline{A}B)$ is high. Thus the 1 output of A3 passes through O2 as a 1 carry input to stage X3.

Finally, consider a fourth example where the inputs are A = low, B = high, and C = low. Again, the input to I1 is low and the output is high. The high input to A5 along with the inverted carry input (inverted from low to high by I2) enables A5. The function $\overline{C}(A\overline{B} + \overline{A}B)$ is high. This high output passes through O3 to represent a 1 condition sum for the stage X2 output. Note that A3 receives a high input from I1 but a low (noninverted) carry input, so that no carry is passed to the stage X3.

Full-Adder Using NOR Logic

The adder circuit shown in Fig. 5-27 is representative of parallel using NOR logic. (Note the similarity to Fig. 5-26.) The circuit logic is identified in the truth table in Table 5-2. Note that when any input is low, the output is high. Conversely, when all inputs are high, the output is low.

To understand the operation of this circuit, first assume that the A and D inputs are high and the C input is low ($AD\overline{C}$). In this condition the A1 output is low. This low input causes the

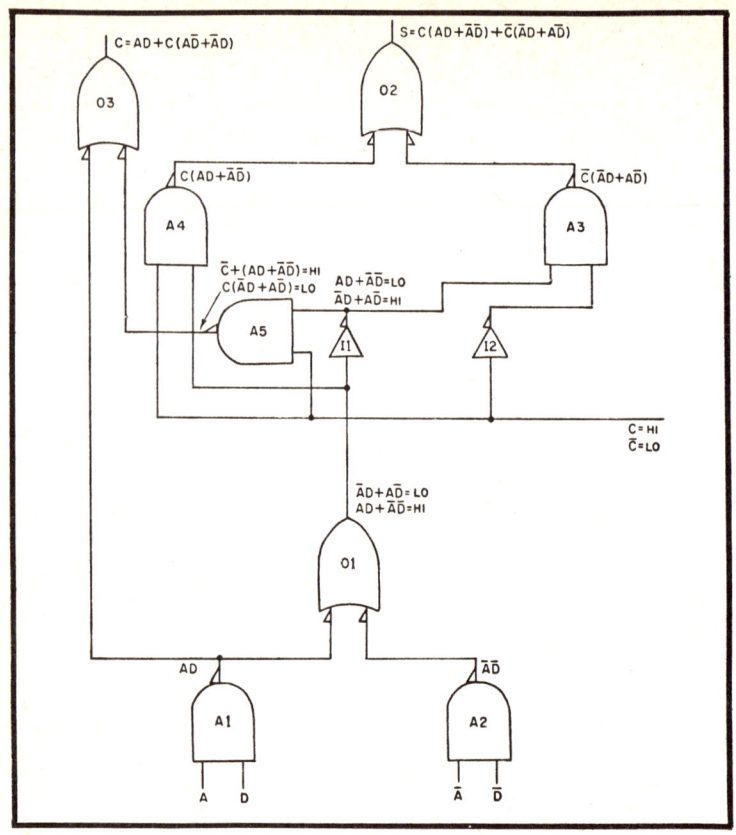

Fig. 5-27. Parallel adder using NOR logic.

Table 5-2. Truth Table for Logic Gates.

A	D	C	LEVEL
Hi	Hi	Hi	Lo
Hi	Hi	Lo	Hi
Hi	Lo	Hi	Hi
Hi	Lo	Lo	Hi
Lo	Hi	Hi	Hi
Lo	Hi	Lo	Hi
Lo	Lo	Hi	Hi
Lo	Lo	Lo	Hi

$AD + \overline{A}\overline{D}$ function at the O1 output to go high. The high O1 output serves as a partial enable to A4. However, the low carry input holds A4 in the disabled condition.

The high O1 output is inverted by I1 so that the $AD + \overline{A}\overline{D}$ function at the I1 output is low to disable A5 and A3. Thus neither A3, A4, nor A5 is enabled to produce a low output. Consequently the O2 output remains low and the sum bit is 0. The low output of A1 and the high output of A5 (A5 not fully enabled) cause the O3 carry output to be high. Thus a carry input is fed to the next higher order stage (not shown).

Now assume that the A, D, and C inputs are all high. The high carry output is generated in the same manner as just described. (The function $AD + \overline{A}\overline{D}$ is low at the I1 output, so that A5 is disabled.) The high output of O1 (derived in the same manner as described in the previous example) is applied to A4 as a partial enable. The high carry input is also applied to A4. Thus the A4 output goes low. This low output is an input to O2 and causes the O2 sum output to become high. Thus, for the assumed condition, both the carry and sum outputs are high (1).

Note the quality of the logic circuits as follows. Two functions are available at each output—the active function and its *complement*. The two outputs are at opposite levels, and obviously cannot occur at the same time. Thus a decision is made when mechanizing the equation as to which function shall be shown in its active state. This is an important consideration in logic mechanization, since either level from a given circuit can be used as the enable to another circuit.

SERIAL AND PARALLEL OPERATION

Adder circuits can be designed to accept input pulses in either serial or parallel form. In parallel operation, each bit of a binary number is carried on a separate transmission line. In serial operation, the binary information is carried in the form of a series of timed pulses. The relative advantages and disadvantages of each form of data transmission are fairly obvious. Parallel operation is much faster, because all bits are transmitted simultaneously, while serial transmission can pass only 1 bit at a time. On the other hand, serial transmission is much cheaper than parallel and requires less equipment.

A true evaluation of the two methods, however, is not quite so simple. For example, because the carry digit must be accumulated in the next higher order, the addition of n number of digits involved in an addition in a parallel machine is not necessarily accomplished n times as fast as can be done by a serial machine. Neither is it true that a parallel machine requires n times as much equipment. Thus it would not be accurate to say that either of the two types of machines has a net advantage over the other, except where all features of the system are known and evaluated.

The choice of serial or parallel operation is also affected by the type of storage and the accessibility of stored data. The time required to read up data from storage and to write information in storage are important considerations which add to the total time required to complete an arithmetic operation in both serial and parallel machines. This time is inherently available in serial machines because of the time between pulses in a train, but must be made available in parallel machines by introducing delay periods.

DIGITAL COMPARATOR

Unlike the radio receiver, a digital system generally exhibits no natural symptoms—such as hum, distortion, or erratic volume—to warn the operator that something has gone wrong. Instead, the system simply provides the wrong answer. Furthermore, component failure is not the only way by which information can be made erroneous or be lost. There is some distortion each time a pulse passes from one circuit to another. Added to this is the distortion caused by the inductive effects from nearby circuits. Thus, it should be understood that a pulse itself can sometimes become so distorted that it may produce an error.

One method of detecting errors, which may be called a *redundancy* method, involves running the problem through the computer twice and noting whether the solutions are identical. However, a computer's major function is to save time, and this type of check doubles the amount of time that is required to solve a problem. In addition, this check is useless if a component has failed; for the faulty component will probably distort both solutions in exactly the same manner.

A better solution is to provide two sets of identical circuitry, as shown in Fig. 5-28. When this is done, the same problem is run simultaneously through both circuits. This is called parallel operation. Unless both circuits commit exactly the same error, obtaining identical solutions indicates that the answer is correct. (You may recall that the exclusive-OR circuit produces an output only when its two inputs are different.)

When using this method, the inputs should always be the same. A signal at the output of any of the exclusive-OR circuits indicates that the two parallel major section outputs are not identical, and that one of them is in error. A signal at *B*, for example, means that either section B1 or section B2 is in error.

Again, the cost factor must be considered. While parallel operation is an excellent technical concept and its use requires no extra time, it does double the circuitry costs. Thus a different solution must be found.

CODES AND WEIGHTING

It is important to understand the *principles* of coding; memorization of the codes is not necessary—in fact, it is sometimes impossible. There are so many different digital codes that no one individual knows all of them. A small portion of such codes is shown below. They are named by their four lowest integral weights, such as 8-4-2-1.

5, 3, 1, 1	6, 3, 1, 1
4, 3, 2, 1	5, 3, 2, 1
4, 2, 2, 1	2, 4, 2, 1
8, 4, 2, 1	6, 4, 2, 1

Digital codes can be divided into three categories:

1—Regularly weighted codes
2—Arbitrarily weighted codes
3—Nonweighted codes

Regularly Weighted Codes

When considering regularly weighted codes, each number position has a weighting value, just as the decimal number

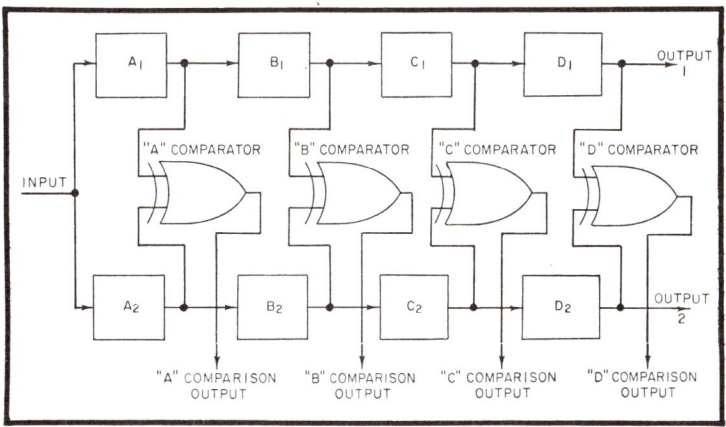

Fig. 5-28. Exclusive-OR circuits used as comparator in error detection.

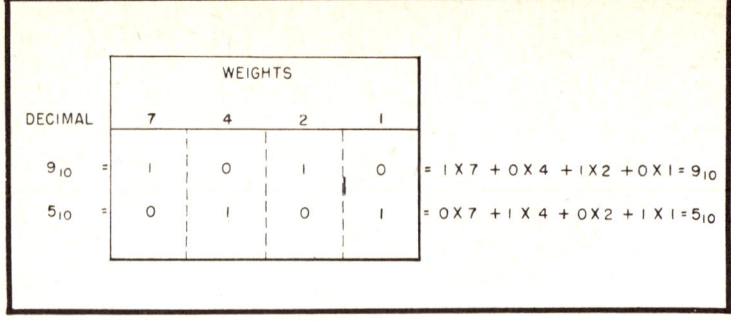

Fig. 5-29. Formation of 9_{10} and 5_{10} in the 7-4-2-1 code.

system, and its weighting values are regular. For example, 9_{10} and 5_{10} are formed in the 7-4-2-1 code as shown in Fig. 5-29. The weighting values form a simple arithmetic series, which begins with 1 and increases regularly.

The important point to remember about regularly weighted codes is that the weights are determined by some rule. If one knows the rule—how the various weights are derived—he can find the weighting value for any desired position.

Arbitrarily Weighted Codes

Arbitrarily weighted codes also use weighting values, but have no rules for forming them. Consider the 4-2-2-1 code, in which the first four weighting values, from left to right, are: 4, 2, 2, 1. There is no rule or formula for generating these values, and one cannot know what the weighting value of the fifth digit from the right might be. However, one will find that the 4-2-2-1 code, as well as many other arbitrarily weighted codes, can be a very practical one, and that it makes no difference what the fifth digit weight might be, simply because it is never used.

Nonweighted Codes

Nonweighted codes have no weighting values at all. Each coded group is defined so as to represent some quantity. The Roman numeral system is such a nonweighted code.

Chapter 6

Elements of Linear Integrated Circuits

Integrated circuits may be classed as either digital or linear. A digital circuit has a discontinuous, or pulsed, output. This makes it useful for applications such as computers, numerical control of machinery, and data processing systems.

Perhaps the best definition of a linear IC is *any IC that is not digital*. Actually, the term *linear* is somewhat misleading as it is applied to ICs. A linear circuit element is one in which a change in the input causes a directly proportional change in the output—double the input, double the output.

But some "linear" ICs do not have such a response—the square-law detector used in radio receivers, for example. Also, such nonlinear circuits as zener diodes are sometimes used in linear ICs.

So, while it is all right to use the term *linear* as it is conventionally used to refer to ICs, it is important not to take it too literally, and to understand its limitations. In any event, the basis of most linear integrated circuits is the linear transistor amplifier, discussed next.

TRANSISTOR LINEAR AMPLIFIER CIRCUITS

In analyzing the qualitative behavior of transistor circuitry, the following generalizations should be very helpful. It should be noted, however, that these generalizations apply only to a transistor circuit that is operated *class-A* (discussed shortly).

The first letter of the transistor type indicates the polarity of the emitter voltage with respect to the base. Thus a PNP transistor has a positive DC voltage applied between emitter and base, and an NPN transistor has a negative DC voltage applied

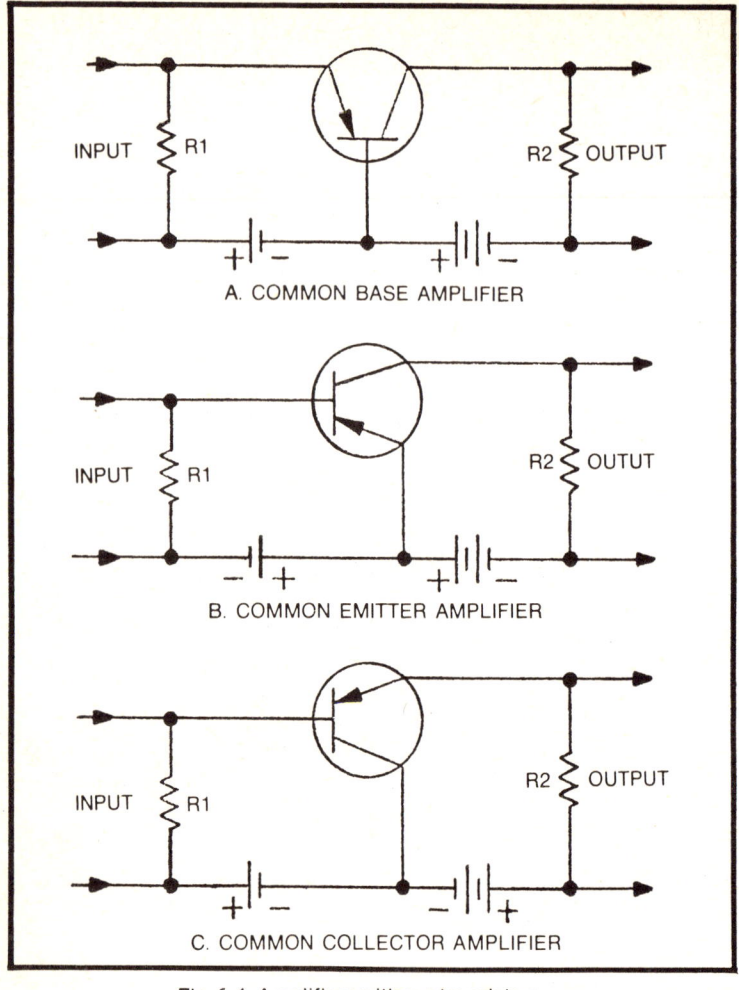

INPUT ⧩ R1

R2 ⧩ OUTPUT

A. COMMON BASE AMPLIFIER

INPUT ⧩ R1

R2 ⧩ OUTUT

B. COMMON EMITTER AMPLIFIER

INPUT ⧩ R1

R2 ⧩ OUTPUT

C. COMMON COLLECTOR AMPLIFIER

Fig. 6-1. Amplifiers with PNP transistors.

between the emitter and the base. The second letter of the transistor type indicates the polarity of the collector with respect to the base. Thus a PNP transistor has a negative DC voltage applied to the collector, and an NPN transistor has positive DC voltage applied to the collector.

From these two facts, you can see that the first and second letters of the transistor type indicates the relative polarities between the emitter and the collector. In a PNP transistor, the emitter is positive with respect to the collector, or the collector is negative with respect to the emitter. In an NPN transistor, the emitter is negative with respect to the collector, or the collector is positive with respect to the emitter.

You will note that the electron current direction is always *against* the direction of the arrow on the emitter. If the electrons flow into the emitter, they flow out of the collector; if they flow out of the emitter, they flow into the collector.

The base—emitter junction is normally biased in the forward direction; the collector—base junction is normally biased in the reverse direction. The reverse direction. The emitter and collector currents will increase with an input voltage that increases the forward bias. Likewise, the emitter and collector currents will decrease with an input voltage that opposes or decreases the forward bias.

Transistor Arrangements

There are several amplifier circuit arrangements possible for the transistor. Figure 6-1A shows the common-base amplifier, in which the signal is introduced into the emitter—base circuit. Since the base element of the PNP transistor is common to both the input circuit and the output circuit, this arrangement is referred to as a *common-base* (CB) amplifier. The common-base circuit is also referred to as the *grounded-base* amplifier.

In the common-emitter amplifier of Fig. 6-1B, the signal is introduced into the base—emitter circuit and taken from the collector—emitter circuit. Here the emitter element of the PNP transistor is common to both the input and the output circuit giving rise to the name *common-emitter* (CE) or *grounded-emitter amplifier*. A common-emitter amplifier using an NPN transistor is similar to the circuit of Fig. 6-1B, but the polarities of the biasing batteries are reversed in order to maintain forward bias in the base—emitter circuit and reverse bias in the collector—base circuit.

Figure 6-1C shows the common-collector amplifier. Because the collector element of the transistor is common to the input and output circuits, the circuit is called a *common-collector* (CC) or *grounded-collector* amplifier. The signal is fed into the base—collector circuit and removed from the emitter—collector circuit.

Regardless of the type of circuit, the proper bias for a class-A transistor amplifier consists of forward bias on the emitter—base circuit and reverse bias on the collector—base circuit. You have just seen in Fig. 6-1 one method of biasing each circuit configuration. Other biasing methods can be used.

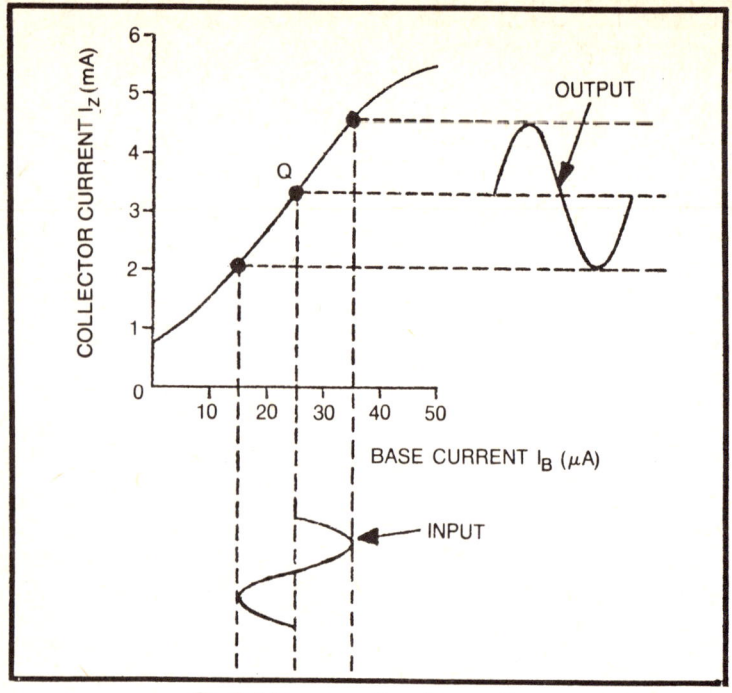

Fig. 6-2. Linear or class A operation.

To study how the collector current of a transistor amplifier acts with a signal current applied to the base, it is helpful to plot the input signal and the resulting collector current along the *dynamic transfer characteristic.* The dynamic transfer characteristic is simply a graph of collector current vs base current for a given load resistance.

Figure 6-2 illustrates linear operation, using a dynamic characteristic curve. The input and output waveforms are drawn on the curve. The output shows no sign of distortion, because the transistor is operating linearly.

The transistor will operate linearly when (1) the proper operating point (point of no signal, Q) is established by the right value of base bias current, and (2) the change of base current is within the linear (straight) portion of the dynamic characteristic curve. Under these two conditions, the amplified output signal will be an exact reproduction of the input signal (Fig. 6-2), which is class-A operation.

If the operating point is either too low or too high, the change in base current will go beyond the linear portion of the dynamic characteristic curve. When the operating point is too

158

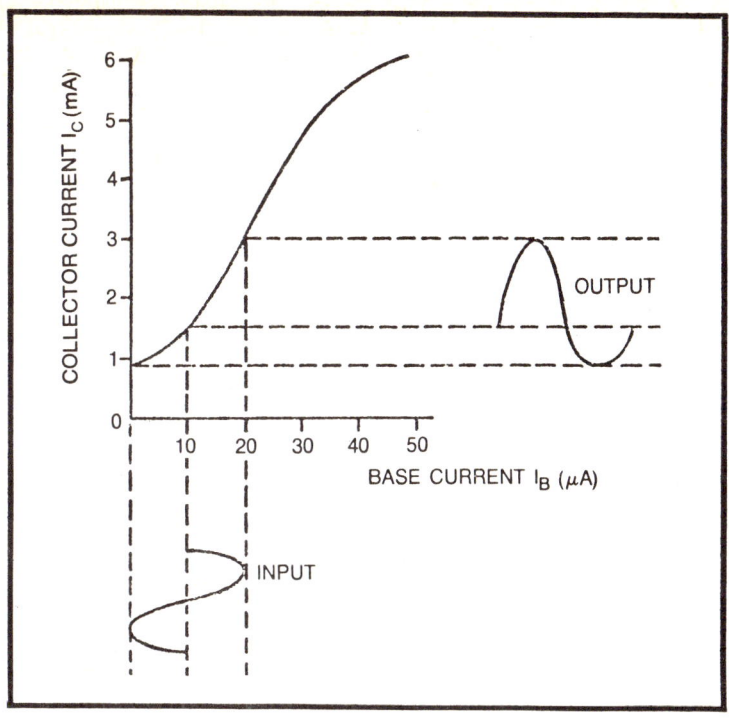

Fig. 6-3. Nonlinear operation.

low, as in Fig. 6-3, the output signal is distorted during the negative change in collector current. When the operating point is too high, the positive-going portion of the input is not on the linear portion of the characteristic curve.

RC-Coupled Amplifiers

RC coupling is used extensively with discrete transistors. In Fig. 6-4, R1, C1, and R2 form the RC network between the two transistor stages. R1 is the collector load resistor for the first stage; C1 blocks DC and couples the AC signal; R2, the DC return resistor for the input element of the second stage, develops the signal applied to the input of the second stage.

C1 prevents the DC voltage of the collector of the first stage from appearing at the input terminal of the second stage. The reactance of the capacitor, which is in series with the input resistance of the following stage, must be small compared to the input resistance. Otherwise, much of the signal voltage will be dropped across the capacitor. A high value of capacitance is used because the input resistance of the following stage is low

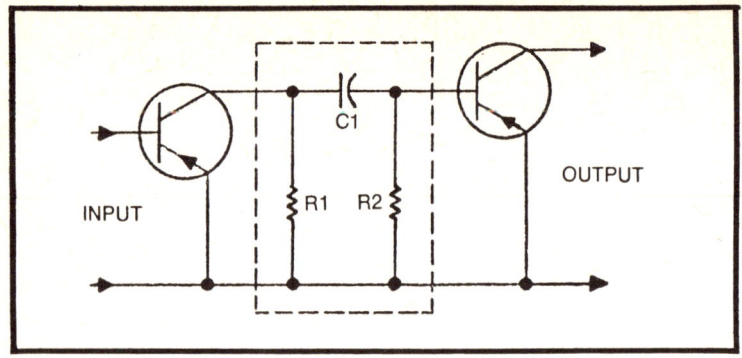

Fig. 6-4. Interstage coupling networks.

(usually lower than 1000 ohms). The ohmic value of the DC return resistor is usually 7 to 15 times the input resistance of the second stage. This ratio of resistance is necessary to prevent shunting the signal current around the input circuit of the second stage.

Since the reactance of the coupling capacitor increases as the frequency decreases, the very low frequencies are attenuated. The shunting effect of the collector—emitter capacitance of the first stage and the base—emitter capacitance of the second stage limits the high-frequency response of the amplifier. The efficiency of the RC-coupled amplifier (ratio of AC power output to DC power delivered to the stage) is low due to the dissipation of DC power in the collector load resistor.

RC coupling is used to a large extent in discrete-device audio amplifiers—from low-level, low-noise preamplifiers to high-level power amplifiers. Other advantages of RC coupling with transistors are high gain, economy of circuit parts, and small size.

This method of coupling is not used in integrated circuits, because it is impossible to fabricate sufficiently large coupling capacitors and it is uneconomic to fabricate sufficiently large DC return resistors using IC technology. Hence, direct coupling is used.

Direct-Coupled Amplifiers

Direct coupling is especially useful for amplification of DC and very low frequencies. It has an economy of parts and lends itself to use in complementary symmetry and integrated circuitry. However, when more than a few such stages are connected in cascade, noise becomes a problem, since the direct-coupled stage does not discriminate against noise. Also, such stages have a rather low efficiency. These disadvantages are outweighed in microelectronics by the relative ease with which such amplifiers can be realized in monolithic ICs. And, of course, the relative inefficiency of the direct-coupled amplifier is less important in IC applications than in high-power applications.

Figure 6-5 shows direct coupling between two stages of audio amplification. The first stage is biased so as to provide a quiescent (no signal) value of 7V on the collector of Q1. Resistor R1 is the base biasing resistor. Resistor R2 serves both as collector load for Q1 and also as the base input resistor for Q2. Resistor R3 is the collector load for Q2. Resistors R4, R5, and R6 form a voltage-divider network to provide the necessary circuit

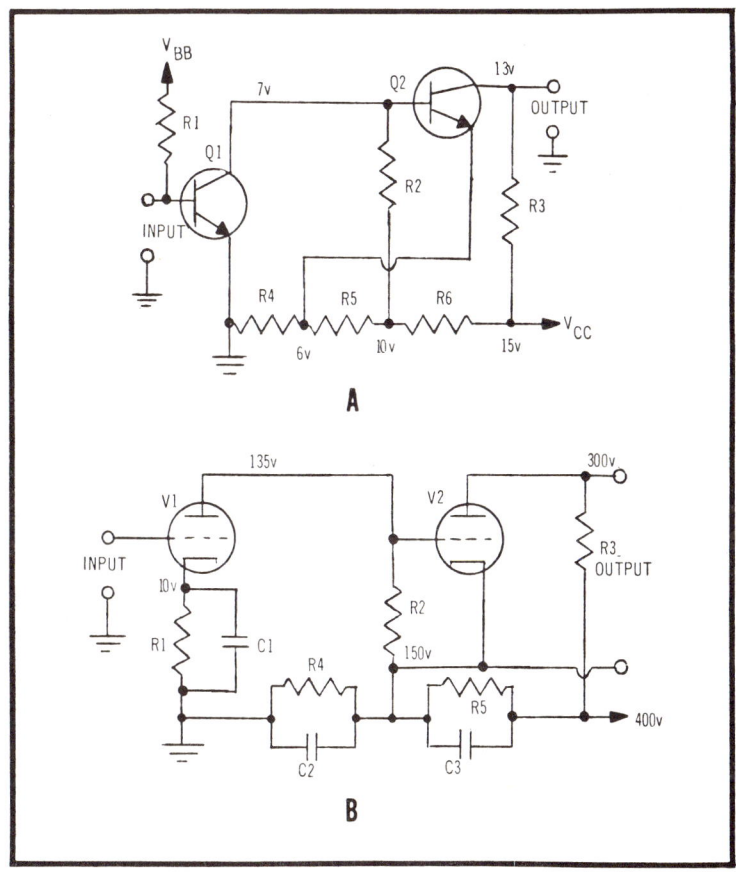

Fig. 6-5. Direct coupling.

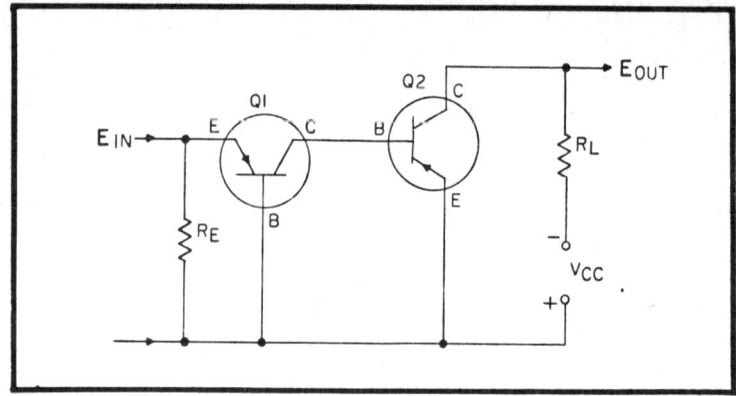

Fig. 6-6. Direct coupling between common-base and common-emitter amplifiers.

voltage requirements. The 7V collector voltage of Q1 is also the base voltage of Q2. The emitter of Q2 is at a potential of 6V, which is developed by the voltage-divider network. This forward biases Q2 and determines its quiescent operating point.

Figure 6-6 shows another basic direct-coupled amplifier connection. The common-base circuit of Q1 is directly connected to the common-emitter circuit of Q2. Thus, the input circuit of Q2 is the load for Q1, and collector bias for Q1 is obtained through the collector—base junction of Q2. Since Q2 biases Q1, only one power source is needed.

A complementary symmetry direct-coupled amplifier is shown in Fig. 6-7. In this circuit, the different natures of the NPN and PNP complement one another. As in Fig. 6-6, the collector bias for Q1 is obtained from the base—collector junction of Q2.

If another stage were added to the circuit of Fig. 6-7, an additional and larger collector bias supply would be required to maintain the negative collector—base potential for each stage. This limitation is analogous to that of the DC supply for the electron tube amplifier. It is evident that a shift of the DC bias potential would be amplified and passed along to the second amplifier, whereas in the AC-coupled (RC-coupled) amplifier, such DC shift would be blocked by the coupling capacitor. Shortly, we shall consider a DC amplifier that overcomes the problem of DC shift.

The use of complementary symmetry, whereby one transistor is used to bias another with DC coupling, affords the minimum of component parts possible, and it represents an economic advantage that is possible only with transistors.

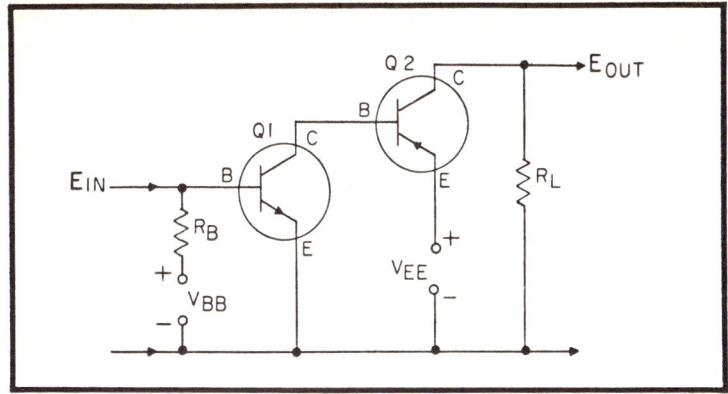

Fig. 6-7. Complementary symmetry amplifier.

DARLINGTON AMPLIFIER

The compound-connected transistor amplifier finds use in the output stage of receivers, public address amplifiers, and modulators; in other words, where large audio power outputs are required. Compound-connected transistors have at least two active elements. A greater number of transistors may be used in the compound connection but we are going to examine only circuits using two compound-connected transistors, the so-called Darlington amplifier. Although two transistors are used, effectively only three leads are used: the base lead of one transistor, the emitter lead of the second transistor, and the collector leads of both transistors which are connected to a common point.

The input to a Darlington power amplifier is a series connection and the output is a parallel connection. To better explain the foregoing statements, we shall use Fig. 6-8 to show the operation of a Darlington audio amplifier circuit as well as its circuit configuration. That part of the circuit enclosed by the dashed lines can be considered as a single transistor with connection points A, B, and C representing the base, collector, and emitter, respectively, of the equivalent transistor.

Fixed class-A bias is supplied to the base of transistor Q1 through voltage-divider resistors R1 and R2, connected across the common power supply. The bias of Q2 is supplied by emitter current flow through Q1 and Q2 to ground. Capacitive input coupling is provided by C1. The output load (R_C) is connected in series with both transistors and the common source of power. The collectors of both transistors are connected in parallel and thus share a common load. The output is taken through coupling

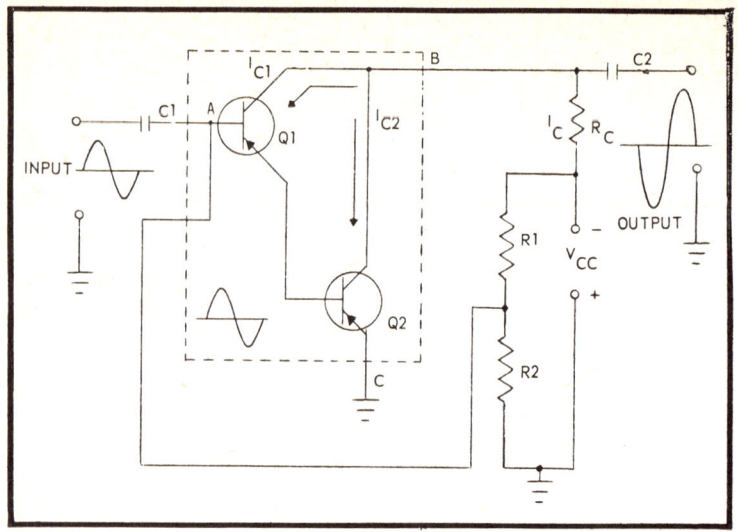

Fig. 6-8. Darlington common-emitter audio amplifier.

capacitor C2. If greater power output is desired, R_C and C2 may be replaced by a transformer.

Both transistors are connected in the common-emitter configuration. The base of Q2, being directly connected to the emitter of Q1, provides a certain portion of negative feedback. Since both transistors are series-connected across the input, a much higher input impedance (50 or more times as large) is offered than with a single common-emitter stage. In the absence of an input signal, both transistors operate in the class-A region. The bias on Q2 is slightly less than that appearing on the base of Q1 because of the small voltage drop between the base and emitter of Q1.

Assuming a negative-going input signal at the base of Q1, we find the collector current increasing due to the increase in forward bias. The amount of increase in collector current is determined by the amplitude of the incoming signal and the gain of the transistor. The incoming signal also forward biases Q2, increasing the collector current. The increase in the collector current of Q2 is not as great as that of Q1, since the forward bias of Q2 is not as large. The collectors, being connected in parallel, have their currents add in the output, since both flow in the same direction through R_C. The increase in current flow through the load will increase the voltage drop across the load and will reduce the voltage on the collectors. Since V_{CC} is a negative voltage in the case of the PNP transistors being used in Fig. 6-8, the output waveform is going in a positive (less negative) direction. The negative-going input signal produces a positive-going output signal.

Conversely, when the input signal is going in a positive direction, it opposes forward bias and reduces the current in the circuit. A reduction in collector current means a lower voltage drop across the load resistor R_C and more voltage on the collector. A positive-going input signal produces a negative going output signal. The characteristic of phase reversal between input and output in common-emitter configurations is still present in the Darlington amplifier.

Current gain in Darlington-connected transistors is greater than that of two transistors connected in cascade by approximately 10%. The addition of reactive elements which affect frequency response in the cascaded amplifier circuit is not present in the Darlington amplifier due to the direct connection of the transistors. Frequency response, therefore, in the Darlington amplifier is not deteriorated.

Although the input impedance in the Darlington amplifier is much higher than that of a single transistor amplifier, the input drive required for operation is the same in both circuits. Remembering that current in a series circuit is uniform and that that a smaller current through a high resistance can produce the same voltage drop as a higher current through a smaller resistance, it can be seen that more drive is not necessary. Some additional gain is due to the two collector currents adding in the output circuitry. The dynamic output impedance of a Darlington amplifier is lower than that of a conventional amplifier, due to increased currents through the load.

The Darlington common-collector amplifier uses the same arrangement of transistors as does the common-emitter amplifier. The major change, of course, is in the output circuit. In the common-emitter amplifier, we take our output from the collector circuit; in the common-collector amplifier, the output is taken from the emitter circuit.

Although the voltage gain of the common-collector amplifier is less than unity, the current gain is high as is the power gain. Figure 6-9 shows a Darlington common-collector amplifier. Note that this circuit uses four leads in this configuration (*B1*, *B2*, *C*, and *E*), although lead *B2* may be omitted.

The Darlington connection can readily be fabricated in integrated circuits and is used to good advantage in many linear ICs.

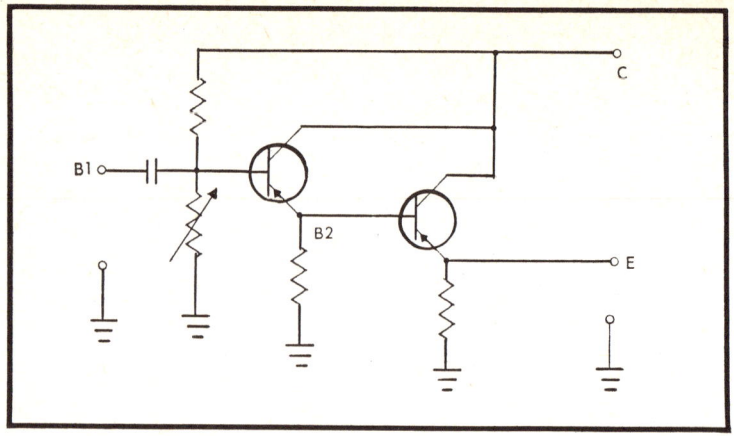

Fig. 6-9. Darlington common-collector amplifier.

DIFFERENTIAL AMPLIFIERS

The *common-mode rejection ratio* (CMRR) is the most significant specification of a differential amplifier (dif-amp) in terms of measurement accuracy. The ability of a dif-amp to reject *common-mode* signals—equal signals appearing at both input terminals—is the figure of merit of the amplifier.

Linear amplifiers, including dif-amps, have one thing in common: gain. That is, an amplifier's output is equal to the gain times the voltage difference between the input terminals, as illustrated in Fig. 6-10. This leads to two important facts:

1—The amplifier output voltage E_0 can be expressed as

$$E_O = A_V(E_1 - E_2)$$

where E_1 and E_2 are the input terminal voltages.

2—The gain A_V of an amplifier is fixed and does not depend on external factors.

A dif-amp is a double-ended amplifier of the push-pull or paraphase type, which amplifies the difference between the two signals applied to its input terminals. If the two signals are identical, there is no difference to be amplified and hence no output. This is illustrated in Fig. 6-10. Assume that the amplifier has a gain of 10 and, as we said, this gain is fixed.

The voltage difference between the input terminals may be a difference in amplitude or phase, or both. It may be described as the graphical addition of the two signals appearing at the input terminals. Referring to Fig. 6-10, we can see that the first pulse (A) is applied only to the *positive* or *noninverting* input of the dif-amp. The output of the dif-amp will then be positive if the input is positive and negative if the input is negative; that is, the positive input is the *in-phase* input of the dif-amp. If the input pulse (E_1) is 1V, then the output (E_0) is 10 times this voltage, or 10V.

The second pulse (B) is applied to the *negative* or *inverting* input. A positive pulse applied to the inverting input will produce an output of opposite polarity, so in this case the output will be negative. The inverting input of the dif-amp is the *out-of-phase* input. If the input pulse is 1V, then the output is -10 times the input, or -10V.

The third pulse (C) illustrates the subtractive nature of the two inputs. Assume that $E_1 = +1$V and $E_2 = -1$V. The difference is then $(+1) - (-1) = +2$V. As a result, the output of the dif-amp would be $+20$V.

The common-mode rejection of the dif-amp is illustrated by the fourth pulse (D). These two input pulses are identical, so their difference is zero. There is no voltage for the dif-amp to amplify. The output of the dif-amp is then zero. Input signals such as signal (D) are termed *common-mode* signals. These are the signals that are rejected by dif-amps.

Whether or not a dif-amp will reject a common-mode signal depends on two things: the electrical parameters of the active devices in the amplifier, and the impedances over which the

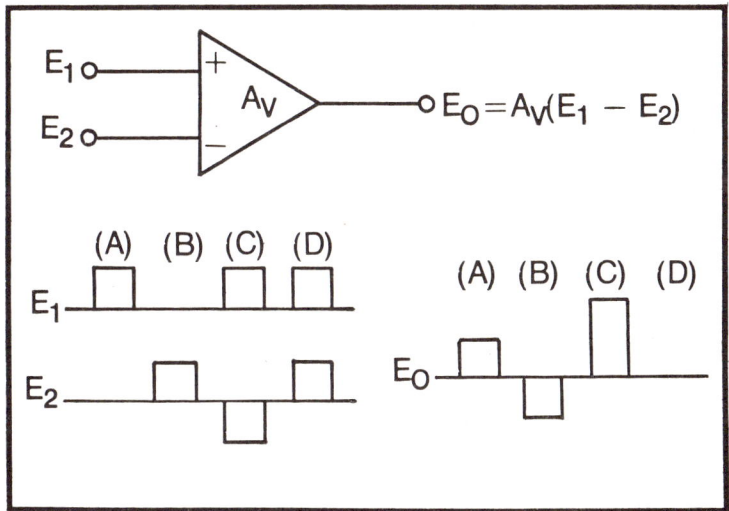

Fig. 6-10. Differential amplifier and signals.

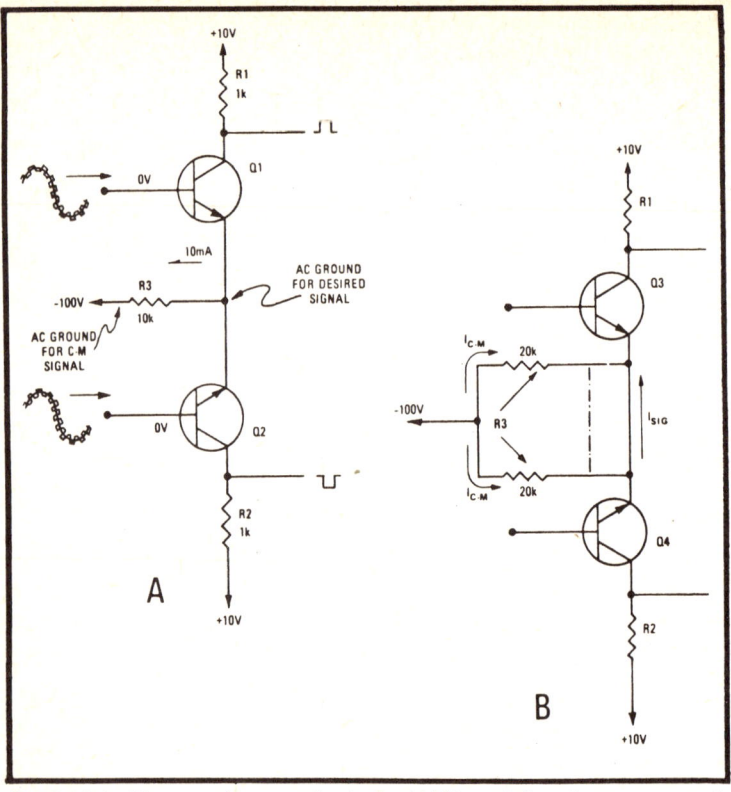

Fig. 6-11. In (A), ground as seen by desired (differential) and common-mode signals. In (B), equivalent circuit and signal paths.

desired signal and the common-mode signal develop with respect to ground points.

The differential and common-mode signals see two different grounds in a dif-amp such as the one in Fig. 6-11A. The desired (differential) signal sees an AC ground between the two emitters, since the input signals to a push-pull amplifier cancel at this point, producing a zero reference point. The common-mode signal, on the other hand, recognizes only the actual chassis ground, the $-V_{CC}$ point.

R3 is the equivalent of two 20K resistors in parallel, as shown in the equivalent circuit, Fig. 6-11B. The common-mode signal will be divided between these two resistors, while the desired (push-pull) signal will bypass the resistors. Thus, the gain in the common-mode signal will be reduced by losses across the 20K resistors, while the gain in the desired signal will be undiminished.

168

The gain in the signal might be calculated as follows, using typical values:

$$A_V = \frac{R1 + R2}{R_t1 + R_t2} = \frac{1000 + 1000}{10.2 + 10.2} = 98$$

where R_t is the dynamic emitter resistance, including the quotient of base-spreading resistance and beta.

The common-mode gain in the sine-wave hum would then be

$$A_V = \frac{R1 + R2}{R_t1 + R_t2 + 2R3} = \frac{1000 + 1000}{10.2 + 10.2 + 2(10,000)} = 0.01$$

The dif-amp in Fig. 6-11A is current-driven by a technique known as "longtailing." The longtail in the circuit is the 10K resistor (R3) which supplies the current for both transistors. Generally, an amplifier current source approaching the ideal current generator allows high CMRR; for example, an infinitely high-impedance longtail returned to an infinitely high supply voltage. The constancy of the current source is one of the most important considerations in dif-amp design. Another important consideration is balancing of the two stages of the push-pull amplifier. Both stages should ideally have the same phase shift and amplification.

If the source impedances are different, as they usually are, the apparent CMRR will be reduced even though the voltages from both sources are the same. This is illustrated in Fig. 6-12. In Fig. 6-12A, two 10.0000V sources designated A and B feed a dif-amp. Source A has an impedance of 100Ω, source B, 50Ω. Because the source impedances are different, the voltages at terminals A and B of the differential amplifier are different. One is 9.9990V and the other is 9.9995V, for a difference of 0.0005V, or 0.5 mV. The apparent CMRR resulting from this is graphed in Fig. 6-13. Notice that the DC CMMR for low frequencies in 20,000:1.

In the case just described, the degradation of CMRR is not too bad. A different situation occurs in the case where the source impedances are relatively high, as in Fig. 6-12B. There the A and B source impedances are 10K and 5K, respectively. The impedance difference results in a differential input difference of 9.95V minus 9.90V, or 0.05V (50 mV). Figure 6-13 shows the effect on CMRR. Note that at low frequencies the CMRR is only 200:1 for the high-impedance sources. Also note the way the CMRR drops with increasing frequency.

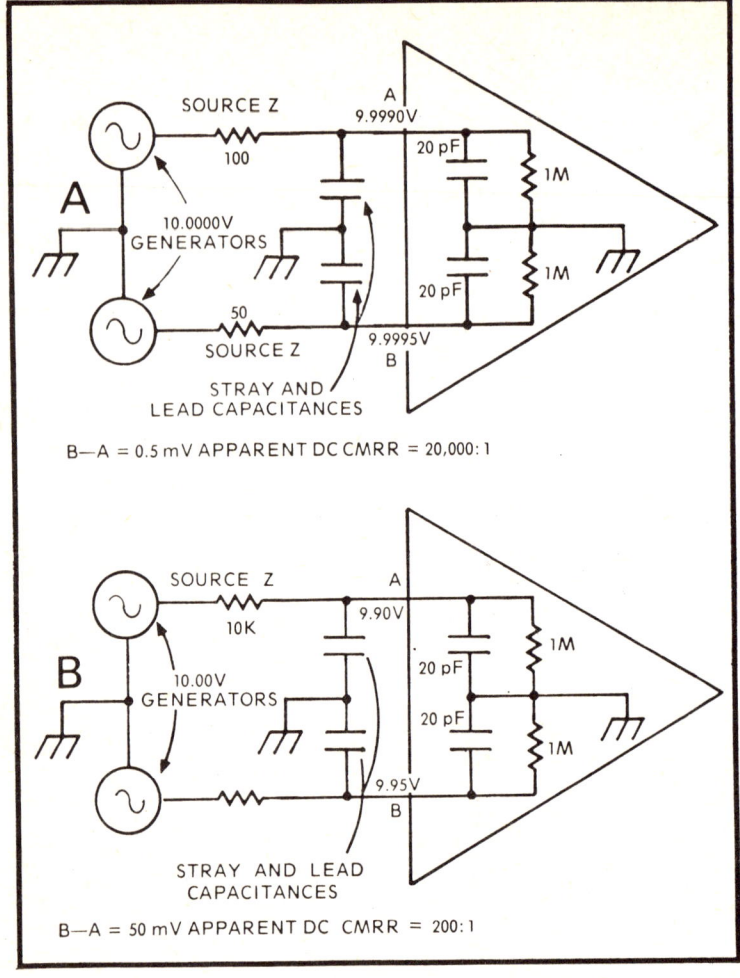

Fig. 6-12. Effect of difference in source impedances on CMRR.

The required balance of the two stages of a differential amplifier circuit is easily achieved in an integrated circuit, since the stages are fabricated at the same time under almost identical conditions (being fabricated very close together). The very close operation also affects each stage equally and the effects tend to cancel.

Negative Feedback

The amplification of an integrated transistor amplifier depends not only on the design of the amplifier itself but on the

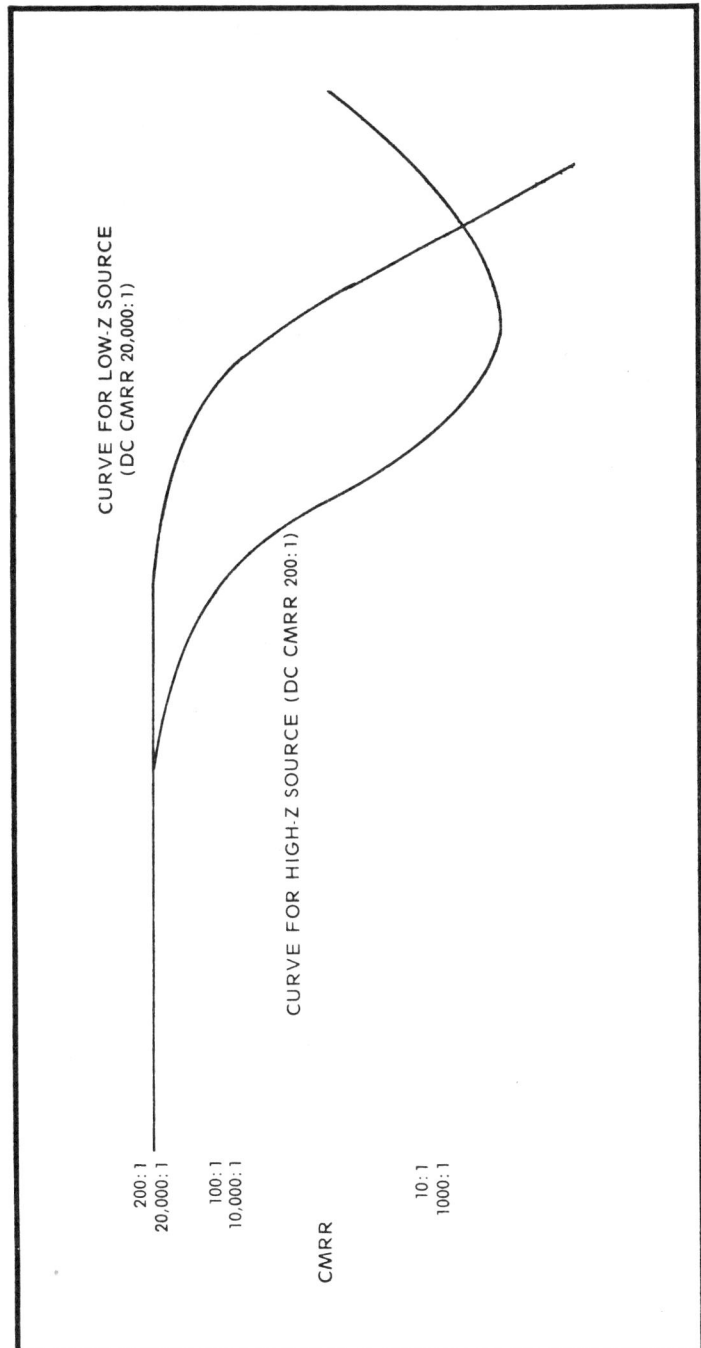

Fig. 6-13. Effect of frequency on CMRR.

CURVE FOR LOW-Z SOURCE
(DC CMRR 20,000:1)

CURVE FOR HIGH-Z SOURCE (DC CMRR 200:1)

CMRR

200:1
20,000:1

100:1
10,000:1

10:1
1000:1

171

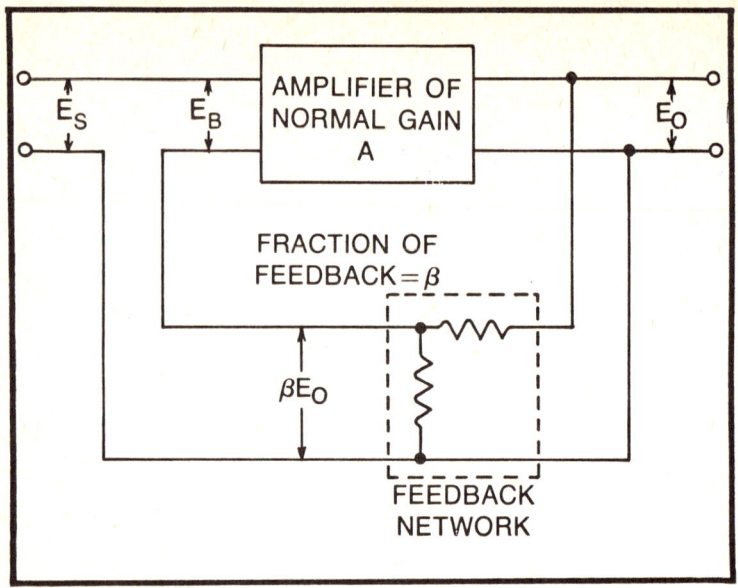

Fig. 6-14. Feedback employing series injection.

supply voltage and ambient temperature as well. Negative
feedback can be used to stabilize the amplification to a precise
value. Assume that a signal voltage (E_S) is applied to the input
terminals, as shown in Fig. 6-14. Let a portion (βE_O) of the
output voltage (E_O) be fed back in series with E_S in such a way
that the signal (E_B) appearing between the base and emitter is
of the form

$$E_B = E_S + \beta E_O \qquad (6\text{-}1)$$

where β is a fractional part of E_O.

Since the normal gain (A) of the amplifier is defined as

$$A = \frac{E_O}{E_B}$$

then by transposing we have

$$E_O = AE_B \qquad (6\text{-}2)$$

Substituting the value of E_B from Eq. 6-1 in Eq. 6-2 we have

$$E_O = A(E_S + \beta E_O)$$

and solving for E_O,

$$E_O = AE_S + \beta E_O$$

$$E_O - \beta AE_O = AE_S$$

$$E_O(1 - \beta A) = AE_S$$

$$E_O = \frac{AE_S}{1 - \beta A} \qquad (6\text{-}3)$$

If βA is greater than 1, the quantity $1 - \beta A$ is negative and the resulting amplification is less than it would be without feedback. Thus the amplification is said to be *negative*, or *degenerative*.

The resulting gain A_R (with feedback considered), is

$$A_R = \frac{E_O}{E_S}$$

Substituting Eq. 6-3 in the above expression,

$$A_R = \frac{AE_S/(1 - \beta A)}{E_S} = \frac{A}{1 - \beta A}$$

The resulting amplifier gain is expressed in terms of the gain without feedback (A) and the fraction of the output (β) fed back to the input. A_R, A, and β may be complex quantities. Generally the feedback factor (βA) is so much larger than 1 that the resulting amplification (A_R) for all practical purposes may be expressed as

$$A_R = -\frac{1}{\beta}$$

Feedback Advantages

Negative feedback may be used to reduce the nonlinear distortion; that is, to make the output waveform more like the input waveform by reducing the nonlinearities that are introduced within the amplifier itself. This use may be understood by the following considerations.

The input signal applied to the base of a transistor amplifier is amplified by an amount determined by the h_{fe} (or beta) of the transistor. Nonlinearities introduced within the transistor are also amplified. If a portion (βA) of the output is fed back 180° out of phase with the input, the distortion component of this inverted feedback voltage will be amplified along with the original input signal and distortion.

The amplified distortion component will tend to cancel the distortion component introduced within the transistor, and the output may be practically free of nonlinear distortion. However, the overall gain of the desired signal will also be reduced; but increasing the number of stages compensates for this reduction.

Noise introduced within an amplifier may be reduced by negative feedback in the same manner that nonlinear distortion is reduced. The same limitations also apply; that is, for feedback to be effective, the noise to be canceled must be generated in a stage around which the feedback is applied.

When it is desired to have the amplification vary in some specific manner with respect to frequency, the negative-feedback network through which β is obtained may be designed to attenuate specific frequencies. For example, if the high frequencies are to be amplified more than the low frequencies, only the low frequencies will be fed back to the input.

The overall gain of a negative-feedback amplifier can be made substantially independent of the magnitude of the load impedance provided that the load impedance does not interfere with the feedback signal. For example, as the effective load resistance is reduced, the AC component of the collector voltage tends to decrease. Accordingly, there is less negative feedback and the amplitude of the base signal is increased, and this offsets the tendency of the output voltage to drop so that the overall gain is nearly constant. On the other hand, if the effective resistance of the load is increased, the AC component of collector voltage will tend to increase and the amount of negative feedback will increase. Thus, the amplitude of the base signal is decreased and the tendency for the output voltage to rise is checked. Again, the overall gain is held nearly constant.

In negative-feedback amplifiers, nonlinear distortion, noise originating within the stage, and frequency distortion may be reduced. In other words, amplitude and phase characteristics can be corrected by negative feedback. Likewise, the effects of variations in loads, supply voltage, and temperature may be effectively counteracted. The price paid for these advantages is a reduction in gain, necessitating an increase in the number of amplification stages.

The principle of response shaping by negative feedback is illustrated in Fig. 6-15.

OPERATIONAL AMPLIFIERS

In its original conception, the operational amplifier (op-amp) was a high-gain DC amplifier whose operational

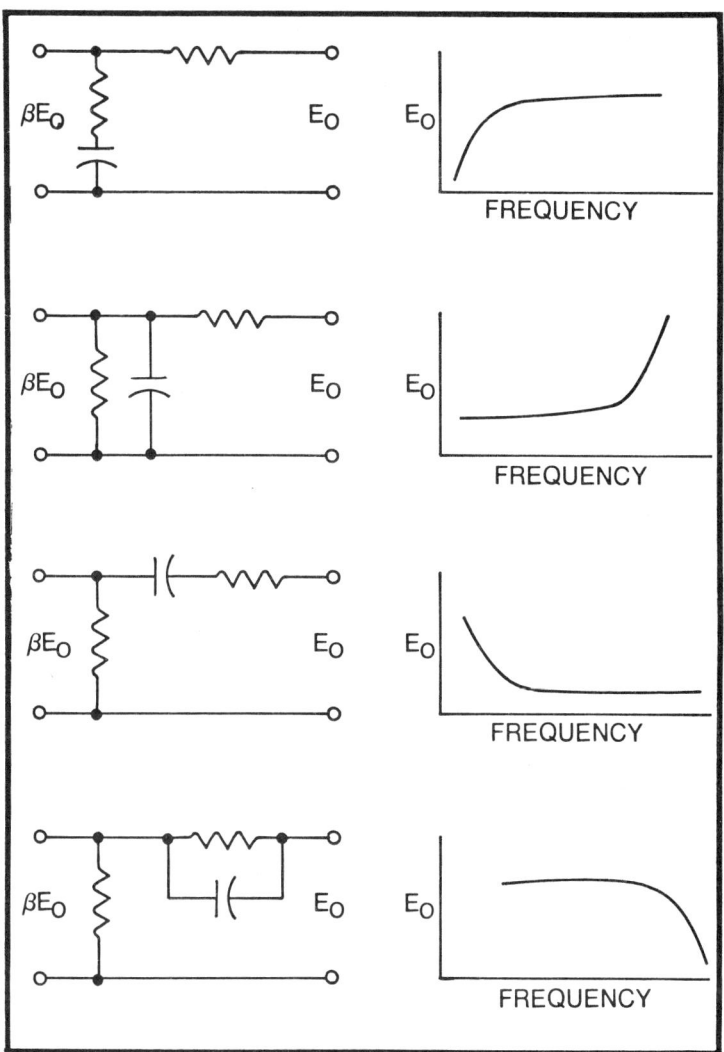

Fig. 6-15. Response shaping by means of negative feedback networks.

behavior could be precisely controlled by the proper selection of components used in its negative-feedback loop. Such amplifiers normally had but one input and one output, meaning that the input was inverting or out of phase with the output. Today, however, the term *op-amp* is generally used to apply to any high-gain DC amplifier having single or differential inputs, single or differential outputs, and whose gain characteristics are controlled by the selection of feedback components.

175

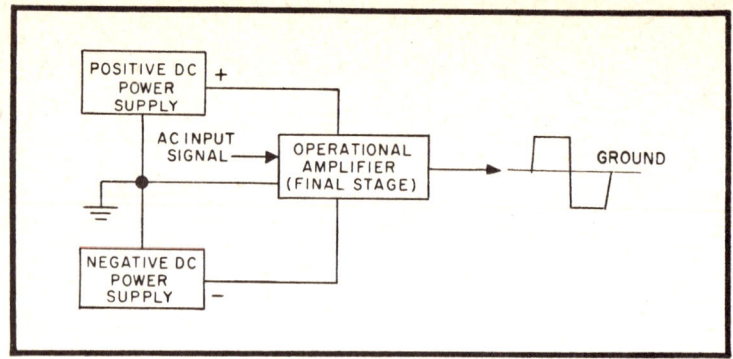

Fig. 6-16. Basic block diagram of operational amplifier, showing method used to obtain positive and negative voltages at the output.

An important feature of op-amps is that the input signal is usually referenced to ground, or zero volts. The input signal is then permitted to swing either above or below ground level, either positive or negative with respect to ground. In order to accomplish this versatile feature the op-amp is powered by two power-supply voltages, one positive and one negative, as illustrated in Fig. 6-16.

Any dif-amp can be used as an operational amplifier. Figure 6-17 shows the typical connections that must be made to the dif-amp. Note in particular that the positive or noninverting input is connected to ground. The op-amp thus amplifies signal voltages that vary with respect to ground. The gain of the op-amp is determined by the components used in the feedback loop. When simple resistive feedback components are used, the op-amp becomes a precision DC amplifier whose gain is equal to the ratio of the feedback resistance (R_F) to the input resistance (R_{IN}); that is, $A_R = R_F/R_{IN}$. And this gain relationship remains

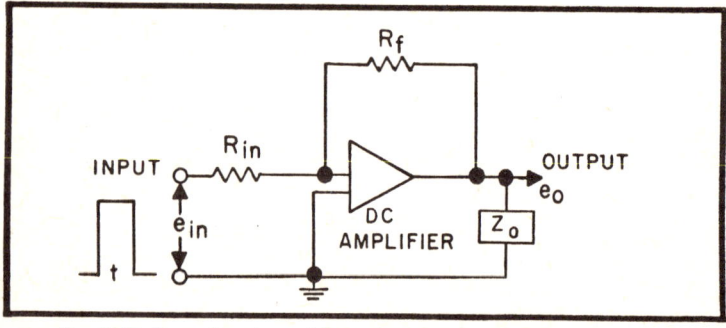

Fig. 6-17. Operational amplifier using resistive feedback circuit.

fairly accurate provided that the *closed-loop* gain (A_R) remains relatively small compared to the *open-loop* or maximum gain (A) of the amplifier without feedback.

The gain of the op-amp can be easily controlled using various combinations of resistances and capacitances. The frequency behavior of the feedback circuits shown in Fig. 6-15 are often used to shape the response of the op-amp. Two very useful op-amp circuits that use such feedback components are the *integrator* and *differentiator*, which are discussed in the following two topics.

Op-amp Integrator

A simple and commonly used integrator consists of two circuit elements: a resistor and capacitor. (See Fig. 6-18.). The voltage across the capacitor is proportional to the integral of the charging current. It can be explained by considering that the voltage across a capacitor is $E = Q/C$. For any given capacitor (C), the voltage depends directly on the charge (Q) which is the imbalance of electrons on the two capacitor plates. The amount of this charge depends both on the amount of current flow and the time for which this flow exists.

The fact that the voltage is proportional to the integral of the charging current allows the RC circuit to be used to make an integrator. The capacitor voltage is the integrator output. Provision is then made to supply a charging current that is proportional to the input signal. The purpose of the resistor is to produce this proportional current from an input signal voltage E_I. At the instant this voltage is applied, the charging current becomes $I = E_I/R$.

Unfortunately, this proportionality does not continue to exist. As the capacitor becomes charged, the capacitor voltage opposes the charging current, and the charging current becomes less proportional to the input signal. This results in an error in the output. The ideal output for a constant input signal is a steadily increasing output. This steady increase is attained only when the signal voltage is first applied and the capacitor has not become appreciably charged.

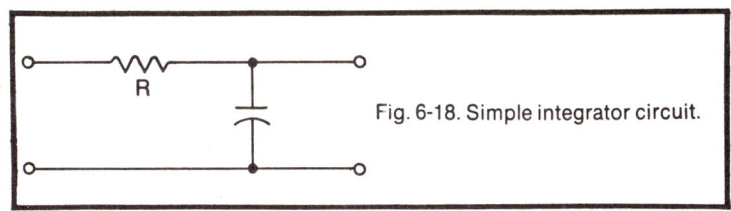

Fig. 6-18. Simple integrator circuit.

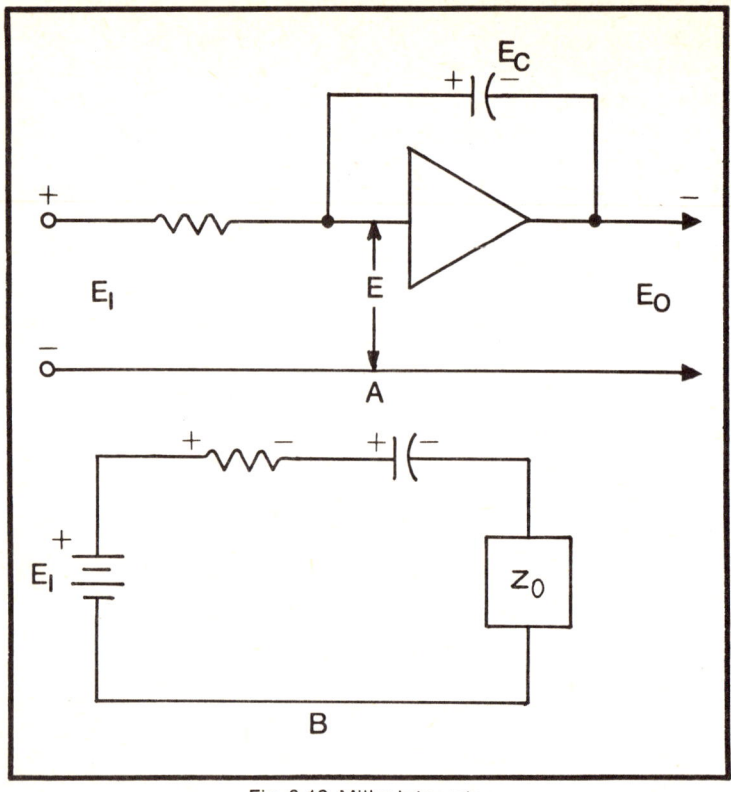

Fig. 6-19. Miller integrator.

A remedy to this error in the RC integrator is to use a circuit with a long time constant, at least with respect to the signal frequencies of interest. Such a circuit delays the charging of the capacitor. The result is a more accurate integration of an input signal. The ideal output response to the square-wave pulse in Fig. 6-17 would be a perfect triangular wave. Although a long time constant produces more accurate results, it also provides a much lower output voltage for the same input signal. Better integration is possible by the use of a high-gain feedback amplifier.

An amplifier integrator is illustrated in Fig. 6-19. This circuit arrangement uses a high-gain amplifier and is known as the *Miller integrator*. The amplifier produces an output which is not limited by the input signal as it is in the simple RC integrator. The amplifier also supplies any energy which is required in the output. The function of the input signal is to control the charging current.

The operation of the Miller integrator can be explained by assuming a constant input as shown in Fig. 6-19A. At the start, assume the initial condition is zero; that is, $E_I = E = E_O = 0$ Also assume that the capacitor is discharged. The positive voltage to be integrated (E_I) is then applied. The capacitor charges with a polarity as shown, since electrons are attracted from the left plate. The charging path is shown in Fig. 6-19B.

A voltage measured at the amplifier input (E) tends to rise in the positive direction since this point is directly coupled to E_I. However, this rise tends to be opposed by the degenerative feedback voltage from the output. The output will be $-AE$ (E_O). The letter A stands for the amplifier gain and the minus sign indicates that the output polarity or phase is opposite to the input. The output changes A times faster or steeper than E. The output voltage is negative and aids the charging of the capacitor.

For a certain input voltage, the charging current is limited to a particular value which tends to keep E practically zero. If the current should exceed this value, E would decrease a small amount due to the increased voltage drop across R. Then E_O would decrease, and the charging current would decrease to the original value. If the initial charging current should decrease, the opposite action would occur. The value of the charging current is therefore stabilized to a specific value proportional to the input voltage. This eliminates the error caused by E_I and the charging current not remaining proportional in the fundamental RC integrator.

This constant charging current must be produced by E_O despite the fact that the steadily increasing capacitor voltage opposes the charging current. To do this, E_O must also steadily increase, and this steady increase in E_O is exactly the integrator output voltage desired for a constant signal input.

Similar action would be produced for a condition in which the input signal suddenly became negative. Polarities then would be in reverse to those shown in Fig. 6-19. Although simple examples are used, the desired result will also be produced for a more complicated signal input. Removal of E_I would produce little effect upon the output which existed at that instant, since the amplifier output would oppose the tendency for the capacitor to discharge.

The limits for E_O are determined by the amplifier and not by E_I or the range of E_C. The output range would be designed to produce an increasing output for any probable input amplitude and period of application. The exception to this would be an integrator which was designed to function also as a limiter.

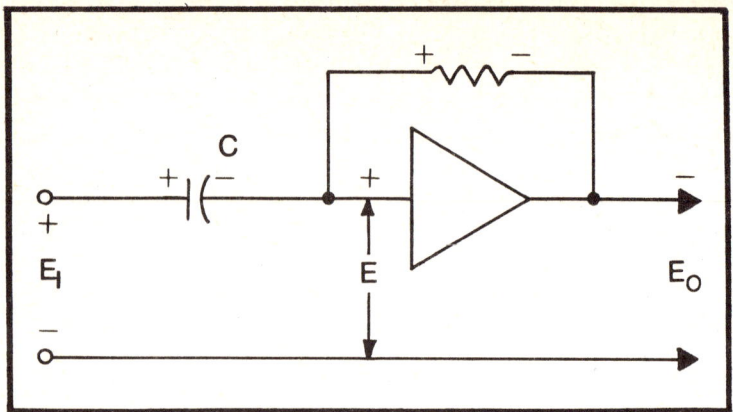

Fig. 6-20. RC op-amp differentiator with positive-going output.

RC Op-Amp Differentiator

As shown in Fig. 6-20, an RC differentiator is just opposite to an integrator. This differentiator circuit has a capacitor input and resistive feedback. The amplifier can be modulator type so as to reduce drift and noise.

In analyzing such a circuit, it is normal to consider the amplifier to have infinite gain. On this basis, E can be considered to be zero. (Although infinite gain is impossible, the amplifier does have a very large gain.) If E_I should change, E tends to change, but the inverse feedback voltage counteracts the change in E. The output voltage depends on the tendency of E to change, and it is proportional to the rate of change of E_I.

This same analysis can be given in terms of polarities. Figure 6-20 shows the circuit assuming a steady, positive input signal. Capacitor C is charged to the steady value of the input signal; therefore, $E = 0$ and $E_O = 0$.

Now, assume the error signal suddenly increases. C begins to lose electrons on the left and gain on the right. The charging of C tends to make E go positive, and this action is reflected in the output as a negative voltage which increases at a rate A times faster than the input. Current produced by the output rapidly charges C to the new value of E_I. Thus, the output is a large pulse of voltage which indicates the rapid change in the input voltage, E_I, and this is the output desired. The amplitude of E_O is not limited by E_I, as was the case in the simple RC circuit, because of the energy supplied by the amplifier. If E_I should decrease, an output pulse of opposite polarity would be produced.

Chapter 7

Digital IC Types

Digital ICs may be categorized according to how their gates are interconnected. They may thus be placed in *logic families*, each logic family having in common a certain type of interconnection and consisting of similar logic functions such as simple gates, inverters, registers, flip-flops, etc.

RESISTOR/TRANSISTOR LOGIC

A resistor/transistor logic (RTL) NOR gate is shown in Fig. 7-1. The logic element is a transistor, and resistors connect it to preceding gates. A positive input at *A*, *B*, or *C* will drive the transistor into conduction, increasing the voltage drop across the collector resistor and decreasing the positive output voltage.

The transistor in this gate is driven by a heavy base current when any input is a logic-1 (positive voltage). In fact, it is made to operate in the *saturation mode*, where further increases in collector voltage would produce no further increase in collector current. Ordinarily, the gate would be made to drive other gates. The number of gates it drives in a given application is the *fan-out*. There is a limit to the fan-out, since the more gates that must be driven, the more current that must be supplied. The fan-out is an important specification of digital ICs. For a given load, a gate with a fan-out of five can drive five circuits with a fan-in load of one, or one circuit with a fan-in load of five.

DIODE/TRANSISTOR LOGIC

The purpose of the input resistors in the foregoing circuit is to isolate the inputs from one another. If diodes are used instead

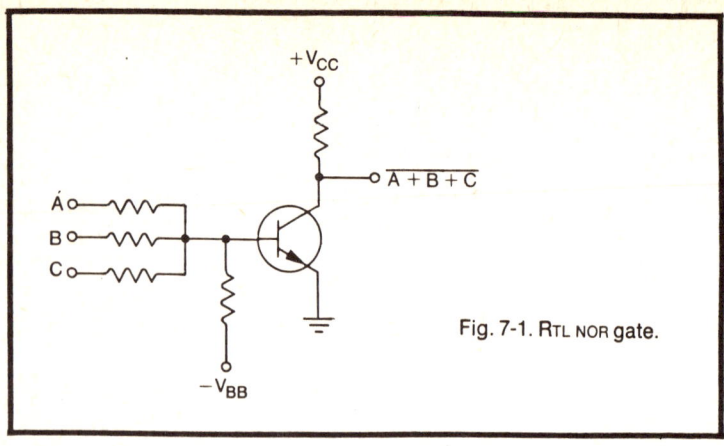

Fig. 7-1. Rᴛʟ ɴᴏʀ gate.

for this purpose (Fig. 7-2), we have diode/transistor logic (DTL). The circuit shown is essentially a diode AND circuit with a transistor inverter to give the NAND function. The diodes provide much better input isolation than resistors. Although the gates in the DTL family can switch states faster than the gates in the RTL family, the diodes place a limit on the switching speed due to the charge stored in their junctions. A logic family designed to get around this problem is discussed next.

TRANSISTOR/TRANSISTOR LOGIC

A transistor/transistor logic (TTL) NAND gate is shown in Fig. 7-3. The multiple-diode arrangement of the DTL circuit is replaced by a multi-emitter transistor. The base—collector junction of this input transistor is never cut off, hence the base

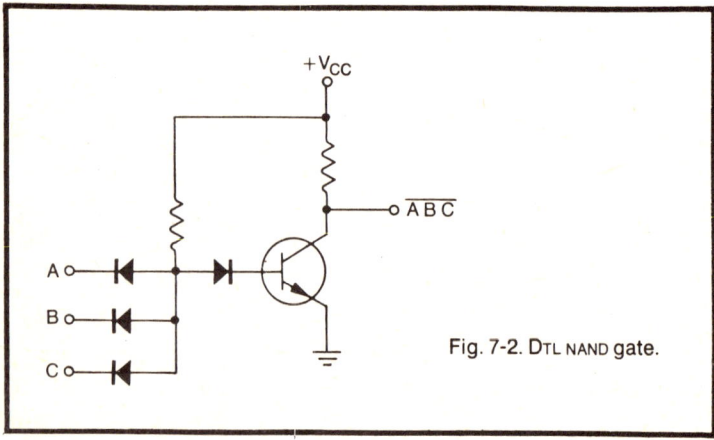

Fig. 7-2. Dᴛʟ ɴᴀɴᴅ gate.

182

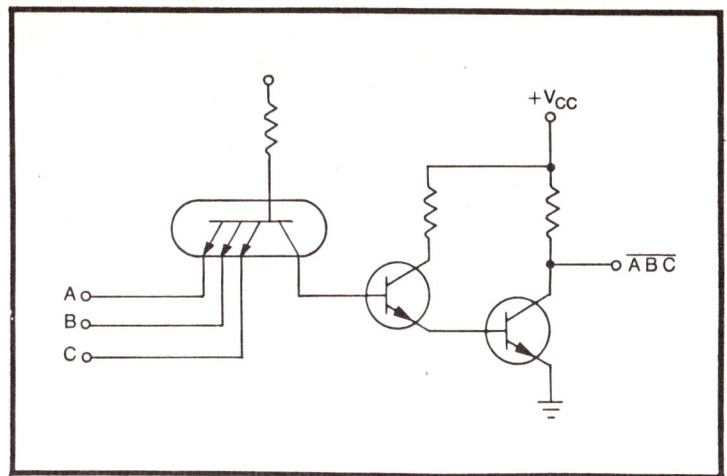

Fig. 7-3. TTTL NAND gate.

current of the next stage always has a low-resistance path. This makes it possible for the base charge to leak off quicker than in the DTL circuit and gives TTL a speed advantage over DTL.

All of the logic families have their advantages, and one is not better in an overall sense than the others. However, the combination of advantages possessed by the TTL family has made it the most popular by far. It offers a high fan-out of 10, fast operating speed, low power dissipation, and good immunity to noise.

Because of its popularity, TTL has become standarized, much as transistors and vacuum tubes before them. It is offered by many manufacturers in a series beginning with the number 5400 or 7400. A 5400 or 7400 IC consists of four 2-input NAND gates, no matter who makes it. Logic gates in this 54/74 series are interchangeable with other ICs of the same number, just as 2N2711 transistors or 6BE6 tubes are. To help you use this all-important logic series, a large listing of TTL ICs is given at the end of this chapter. The list is by no means exhaustive, but it includes the information you will need to use intelligently most of the digital ICs you will come across.

The difference between the 74 and 54 series is one of permissible operating temperatures. The 54 series, or military type, is for use in relatively extreme temperatures; it has a rated operational temperature range of $-55°C$ to $+125°C$. The 74, or industiral, series is for less demanding applications and has a temperature range of $0°C$ to $70°C$. The 74 series is adequate for most purposes.

An improved form of TTL is called *Schottky* TTL. It differs from conventional TTL in that Schottky barrier diodes with metal−semiconductor junctions are used in its construction. Logic gates of this family are designated as the 54S/74S series (beginning with 54S00 and 74S00) and offer much faster switching speeds. Outwardly there is little difference between the 54/74 series and the 54S/74S series since they have identical packaging and pin connections.

Low-power versions of both conventional and Schottky TTL gates are also available; these are designated as the 54L/74L series and the 54LS/74LS series, respectively. Reducing the power consumption of the gates has advantages in reducing the size of the power supplies, but it also slows down the switching speed of the logic gates. It is interesting to note, however, that the low-power Schottky TTL series is slightly faster than the conventional 54/74 TTL series although it consumes only about one-sixth as much power.

Emitter-Coupled Logic

Emitter-coupled logic (ECL) is the fastest of all logic families. Its application is limited primarily to large computers, where extreme speed is required. Another advantage is that a complementary output is available ($A + B$ and $\overline{A + B}$, for example) as shown in Fig. 7-4.

This logic form is sometimes referred to as *current mode logic* (CML). When a transistor operates in its nonsaturating

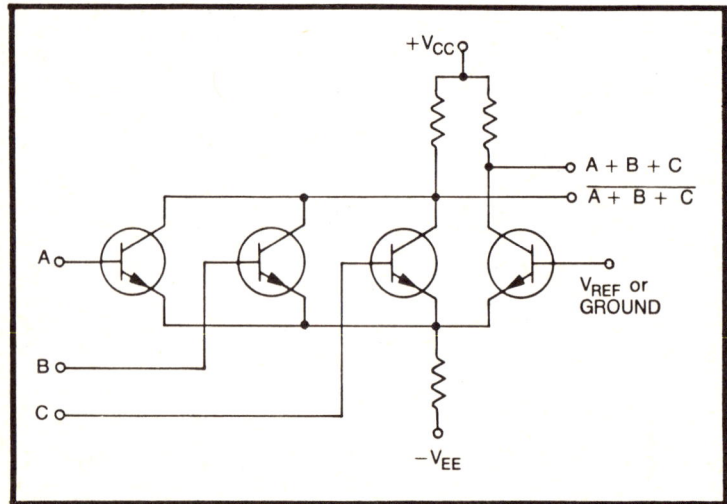

Fig. 7-4. ECL gate showing complementary OR/NOR outputs.

mode, as in these circuits, it is said to be in the current mode. A result of operating in the nonsaturating mode is that the long turnoff delay due to excess charge storage is avoided. This is what gives ECL its high speed.

Complementary MOS Logic

Logic gates can also be made using metal oxide semiconductor (MOS) transistors which can be fabricated with either N or P channels. Of particular interest is the use of both N-channel and P-channel MOS transistors connected in a complementary arrangement. A Complementary MOS (CMOS) logic gate, such as the NOR gate in Fig. 7-5, exhibits extremely low power dissipation. The logic gate is constructed so that when the P-channel transistors (upper transistors in Fig. 7-5) are turned on, the N-channel transistors are turned off, and vice versa. As a result, there is no direct path for current to flow from the power supply to ground except for very small leakage current through the *off* transistors.

Logic gates are constructed in CMOS devices by connecting N- and P-channel transistors in series and parallel. In Fig. 7-5, for example, the NOR logic function is obtained by connecting the upper P-channel transistors in series and the lower N-channel transistors in parallel. A NAND gate could be obtained

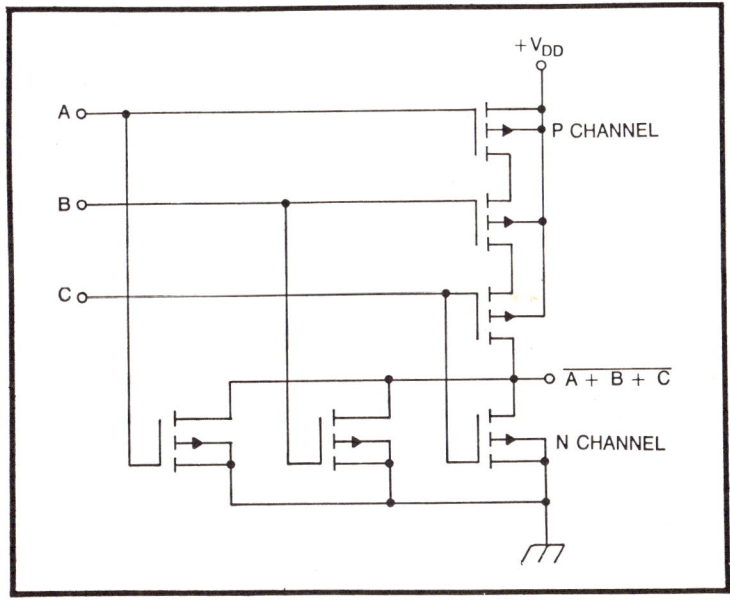

Fig. 7-5. CMOS NOR gate.

Table 7-1. Typical Specifications for Various Logic Families.

Family	Power Supply (volts)	Power Dissipation (mW)	Typical Delay (nsec)	Noise Immunity
RTL	3 ±10% 3.6 ±10%	12	25	Fair
mW RTL	3 ±10% 3.6 ±10%	2.5	45	Fair
DTL	4 ±10%	8	30	Good
VTL	±4 to ±10	12–80	55	Excellent
ECL	5.2 ±20%	25	2	Fair
TTL	5 ±10%	15	10	Good
HTL	12 or 15	30	85	Excellent
CMOS	3–15	0.01	200	Excellent

in a similar manner by reversing the series and parallel arrangement.

CMOS logic uses *enhancement mode* MOS transistors that are normally off and which must be turned on by the application of an input gate signal. The P-channel transistor is turned on by applying a low or logic-0 input level, and the N-channel transistor is turned on by applying a high or logic-1 signal. Complementary operation results from tying N- and P-channel gates together.

In addition to the low power dissipation that results from complementary operation, the CMOS logic gate is not dependent upon a specific power supply voltage; typical devices operate with supply voltages anywhere in the range between 3 and 15V. A disadvantage to MOS transistors is that they switch rather slowly, so CMOS would not be considered a high-speed logic family.

Comparing Logic Families

There are a number of logic families besides those discussed he. Among others, there are RCTL, DCTL, and CTL; but these have largely passed into history. The main ones used today are TTL, for most applications; ECL, for high-performance applications; and CMOS, for low power consumption. Table 7-1 compares some of the features of various logic families. Of the two main modern families, ECL is the fastest and TTL is second. But, in its favor, CMOS has lower power dissipation, higher fan-out, and better noise immunity.

DIGITAL IC SPECIFICATIONS

The following pages list the functional and electrical characteristics of the popular 54/74 series of digital ICs.

Functional specifications tell what an IC does, what decisions it makes. These specifications are often expressed with the help of a truth table or formula.

Remember that positive logic is where logic 1 is positive with respect to logic 0; negative logic, where logic 1 is negative with respect to logic 0. Of course, which type of logic is to be employed is the designer's choice. This choice will affect the very nature of some logic circuits. For example, the NAND gates of the type 7400 IC become NOR gates if negative logic is employed. One may assume that positive logic is employed, however, unless something is said to the contrary. Generally, as here, the logic gates are identified in terms of positive logic. Thus, the 7400 is referred to as a quad 2-input NAND gate. In some publications, you may see the same gate referred to as a NAND/NOR gate. The gate may assume either function, and the author is covering both cases.

The functional specifications given here include information on the applications and special features of the various circuits. For a complete understanding of these applications and features, refer to the appropriate discussions elsewhere in this text.

The electrical specifications include loading information, power dissipation, and speed. The figures given are typical, under normal operating conditions. The loading data, as has already been explained, tells how many circuits of a certain kind may be driven by another given circuit. Unless otherwise noted, all TTL gate outputs can drive up to 10 inputs, a total input load current of 16 mA.

For most semiconductors, the power rating tells how much power the device may absorb as heat without malfunctioning. For ICs, the power rating simply tells how much power the device consumes. We are interested in this primarily because it tells us how much power our power supply must provide. The power consumed by most of these devices is small, but in digital systems where large numbers of ICs are used, the power requirements can become appreciable. The power figures given here tell the typical power usage per package.

The operating speed of most logic devices is expressed as a *propagation delay time*: t_{PHL} is the time required for a decision to change from high to low, and t_{PLH} is the time required for a decision to change from low to high. For all *sequential circuits*, such as flip-flops, counters, and registers, the delay is the time from a clock pulse to a response in the Q or $\bar{Q}$ output.

For sequential circuits, we are frequently more interested in the maximum *toggle rate*. This is the maximum clock frequency that can be used with a sequential device, and as you will see, it is expressed in megahertz.

Positive supply voltage is applied to pin 7 and ground to pin 14 of most 14-pin TTL ICs; pins 8 and 16 are used on 16-pin ICs. Exceptions to this general rule are noted on the data sheets. For power, you can use a +5V supply, but three 1.5V flashlight batteries connected in series to make +4.5V will work just as well for experimenting. Never use four batteries when experimenting, since that would produce +6V which exceeds the maximum +5.5V rating. Higher voltages may be applied to the collectors of some buffer gates having open-collector outputs (see data sheets).

7400
QUAD 2-INPUT NAND GATE

This device consists of four 2-input NAND gates. Each gate may be used as an inverter, or two gates may be cross-coupled to form bistable circuits.

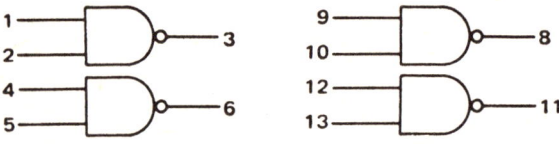

Positive logic:	NAND
Negative logic:	NOR
Total power dissipation:	40 mW
Propagation delay time:	13 nsec.

7401
QUAD 2-INPUT NAND GATE

(Open Collector)

This device consists of four 2-input NAND gates with no output pullup circuits. It can be used where the wired-OR function is required, or for driving discrete components.

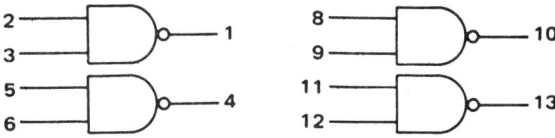

Positive logic:	NAND
Negative logic:	NOR
Total power dissipation:	40 mW
Propagation delay time:	35 nsec
Maximum open-collector output voltage:	5.5V
Maximum open-collector output current:	16 mA

7402
QUAD 2-INPUT NOR GATE

This device consists of four 2-input NOR gates. Each gate may also be used to make an inverter, or two gates may be cross-coupled to form bistable circuits.

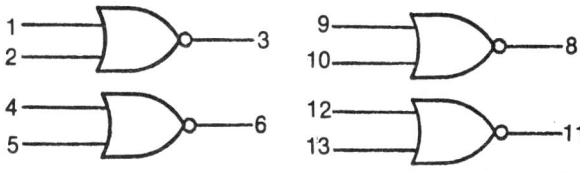

Positive logic:	NOR
Negative logic:	NAND
Total power dissipation:	60 mW
Propagation delay time:	13 nsec

7404
HEX INVERTER

This device offers six independent inverting gates in a single package. Each gate consists of a single input driving an output inverter.

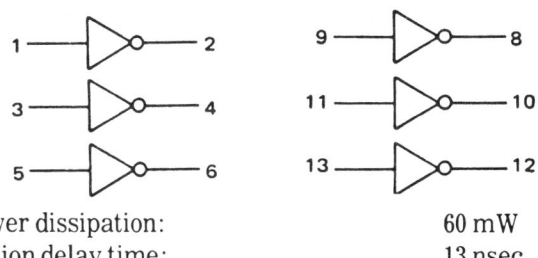

Total power dissipation:	60 mW
Propagation delay time:	13 nsec

7406, 7416

HEX INVERTER BUFFER/DRIVER

(Open-Collector High-Voltage Outputs)

The 7406 and 7416 hex inverter buffer drivers feature standard TTL inputs with inverted high-voltage, high-current, open-collector outputs for interface with MOS logic, lamps, or relays.

Total power dissipation:	160 mW
Propagation delay time, t_{PLH}:	10 nsec
Propagation delay time, t_{PHL}:	14 nsec
Maximum open-collector output voltage:	15V
Maximum open-collector output current:	40 mA

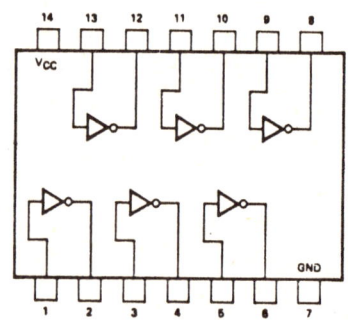

7407, 7417

HEX BUFFER/DRIVER

(Open-Collector High-Voltage Outputs)

The 7407 and 7417 hex buffer/driver features standard TTL inputs with noninverted high-voltage, high-current open-collector outputs for interface with MOS logic, lamps, and relays.

Total power dissipation:	120 mW
Propagation delay time, t_{PLH}:	6 nsec
Propagation delay time, t_{PHL}:	20 nsec
Maximum open-collector output voltage:	30V (7407)
	15V (7417)
Maximum open-collector output current:	40 mA

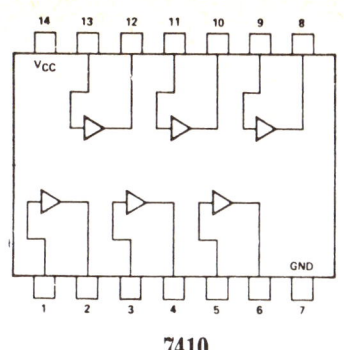

7410

TRIPLE 3-INPUT NAND GATE

Positive logic:	NAND
Negative logic:	NOR
Total power dissipation:	30 mW
Propagation delay time:	10 nsec

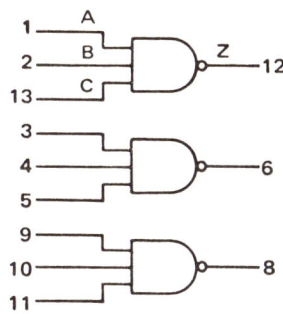

7413

DUAL NAND SCHMITT TRIGGER

This consists of two identical Schmitt trigger circuits in monolithic IC form. Each circuit functions as a 4-input NAND gate, but because of the Schmitt action, the gate has different input threshold levels for positive- and negative-going signals. The *hysteresis*, or *back-lash*, is the difference between the two threshold levels and is typically 900 mV.

An important design feature of the 7413 is built-in temperature compensation. This insures very high stability of the threshold levels and the hysteresis over a very wide temperature range. Typically, the hysteresis changes by 3% over the temperature range of −55°C to 125°C, and the upper threshold changes by 1% over the same range. The 7413 can be triggered from the slowest of input ramps and still give clean, jitter-free output signals.

Total power dissipation:	80 mW
Propagation delay time, t_{PLH}:	18 nsec
Propagation delay time, t_{PHL}:	15 nsec

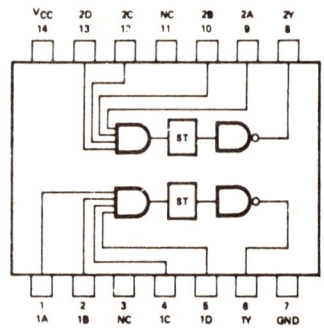

7420

DUAL 4-INPUT NAND GATE

This device consists of two 4-input NAND gates. These gates may be cross-coupled to form an R-S flip-flop.

Positive logic:	NAND
Negative logic:	NOR
Total power dissipation:	20 mW
Propagation delay time:	13 nsec

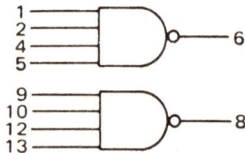

7426

QUAD 2-INPUT NAND BUFFER

(Open-Collector High-Voltage Output)

The 7426 quad 2-input NAND gate features standard TTL inputs with high-voltage open-collector outputs for interface with MOS logic, lamps, and relays.

Total power dissipation:	40 mW
Propagation delay time, t_{PLH}:	16 nsec
Propagation delay time, t_{PHL}:	11 nsec
Maximum open-collector output voltage:	15V
Maximum open-collector output current:	16 mA

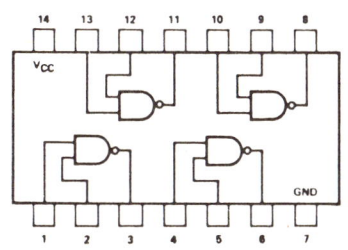

7430
8-INPUT NAND GATE

This device is an 8-input NAND gate. It is useful when processing a large number of variables, such as in encoders and decoders.

Positive logic:	NAND
Negative logic:	NOR
Total power dissipation:	10 mW
Propagation delay time:	13 nsec

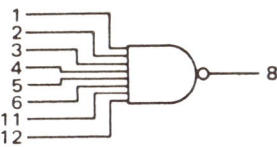

7437, 7438
QUAD 2-INPUT NAND BUFFER

(Open-Collector Output)

The 7437 is a NAND gate similar to the 7400 except that it will drive three times as many loads. The 7438 is an open-collector type, similar to 7403.

The 7437 and 7438 contain four 2-input NAND gates. The 7437 has a guaranteed fan-out of 30 loads. The 7438 has an open-collector output for wired-AND applications, but still retains the high sink-current capability of the 7437.

Total power dissipation:	100 mW
Propagation delay time, t_{PLH}:	10 nsec
Propagation delay time, t_{PHL}:	14 nsec
Output loading factor:	30 (7437)
Maximum open-collector output voltage:	5.5V (7438)
Maximum open-collector output current:	48 mA (7438)

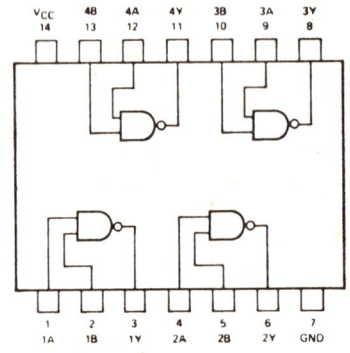

7440

DUAL 4-INPUT NAND BUFFER

This device consists of two 4-input NAND power gate circuits. Each gate is designed for driving 30 input loads.

Positive logic:	NAND
Negative logic:	NOR
Output loading factor:	30
Total power dissipation:	50 mW
Propagation delay time:	13 ns

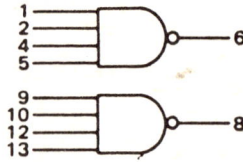

7441

BCD-TO-DECIMAL DECODER/DRIVER

The 7441 Nixie decoder/driver is a 1-out-of-10 decoder which has been designed to provide the necessary high-voltage characteristics required for driving gas-filled cold-cathode indicator tubes. It may also be utilized in driving relays or other high-voltage interface circuitry.

The element is designed using TTL techniques and is therefore completely compatible with DTL and TTL elements. The specially designed output drivers provide the necessary stable output state. There are no input codes where all outputs are off, or where more than one output can be turned on.

Total power dissipation:	80 mW
Maximum open-collector output voltage:	60V
Maximum open-collector output current:	7 mA

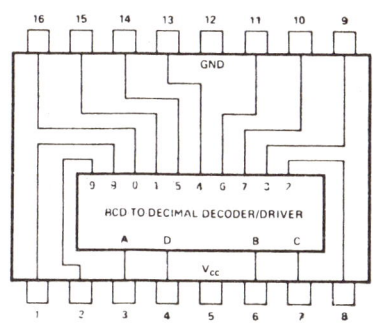

7442

BCD-TO-DECIMAL DECODER

The 7442 BCD-to-decimal decoder is a TTL MSI array utilized in decoding and logic conversion applications. The 7442 decodes a 4-bit BCD number into 1 of 10 outputs.

Total power dissipation:	340 mW
Propagation delay time:	20 nsec
Maximum open-collector output voltage:	55V
Maximum open-collector output current:	7 mA

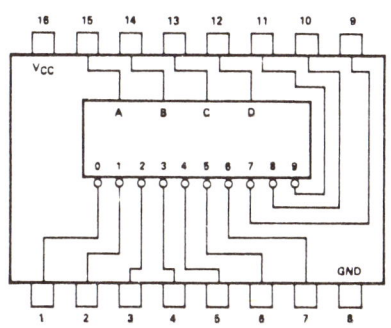

7445, 74145
BCD-TO-DECIMAL DECODER/DRIVER

The 7445 and 74145 BCD-to-decimal decoder/driver is a TTL MSI array. It features standard TTL inputs and high-voltage, high-current outputs.

Total power dissipation:	215 mW
Propagation delay time:	60 nsec
Maximum open-collector output voltage:	30V (7445)
	15V (74145)
Maximum open-collector output current:	80 mA

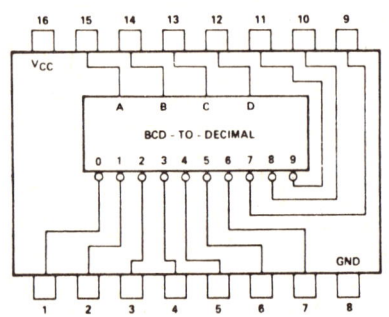

7446, 7447
BCD-TO-7-SEGMENT DECODER/DRIVER

The 7446 and 7447 BCD-to-7-segment decoder driver ICs are TTL monolithic devices consisting of the necessary logic to decode a BCD code to 7-segment readout plus selected signs.

Incorporated in this device is a blanking circuit allowing leading- and trailing-zero suppression. Also included is a lamp test control to turn on all segments.

The 7446 and 7447 provide open-collector output transistors for directly driving lamps.

Total power dissipation:	320 mW
Propagation delay time:	45 nsec
Current per package:	43 mA
Maximum open-collector output voltage:	30V (7446)
	15V (7447)
Maximum open-collector output current:	40 mA

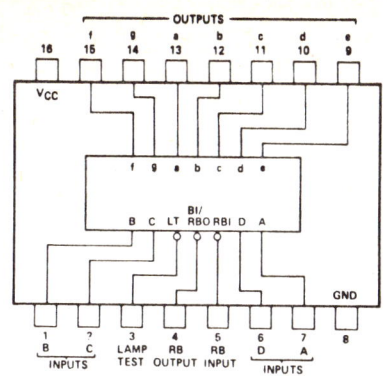

7448
BCD-TO-7-SEGMENT DECODER/DRIVER

The 7448 BCD-to-7-segment decoder/driver is a TTL monolithic device consisting of the necessary logic to decode a BCD code to 7-segment readout plus selected signs.

Incorporated in this device is a blanking circuit allowing leading- and trailing-zero suppression. Also included is a lamp test control to turn on all segments.

The 7448 has a 2K pullup resistor on the outputs to provide sufficient source current to drive interface elements.

Total power dissipation:	265 mW
Propagation delay time:	45 nsec
Maximum low-level output current:	6.4 mA

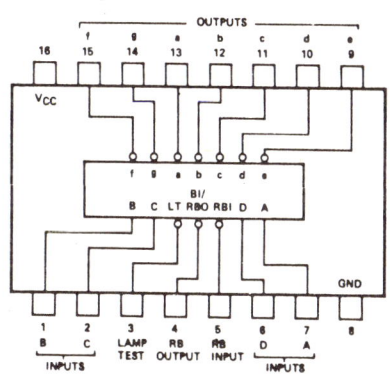

7450
EXPANDABLE DUAL 2-WIDE 2-INPUT AND-OR-INVERT GATE

This device consists of two AND-OR-invert gates, one of which is OR expandable. Each gate is made up of two 2-input

AND gates ORed together and inverted. Up to four 7460 expander gates may be ORed with the device at the expander points.

Positive logic:
$(A \cdot B) + (C \cdot D) + (\text{expanders})$

Negative logic:
$(A + B) \cdot (C + D) \cdot (\text{expanders})$

Total power dissipation: 28 mW
Propagation delay time: 13 nsec

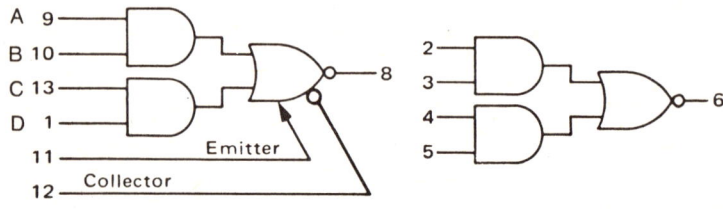

7470
J-K FLIP-FLOP

The 7470 is an edge-triggered J-K flip-flop, featuring gated inputs, direct clear and preset inputs, and complementary Q and $\bar{Q}$ outputs. Input information is transferred to the outputs on the positive edge of the clock pulse.

Direct-coupled clock triggering occurs at a specific voltage level of the clock pulse. After the clock input threshold voltage has been passed, the gated inputs are locked out.

This flip-flop is ideally suited for medium- and high-speed applications. It can be used to obtain a significant saving in system power dissipation and package count where input gating is required.

Total power dissipation: 65 mW
Operating Frequency: 35 MHz

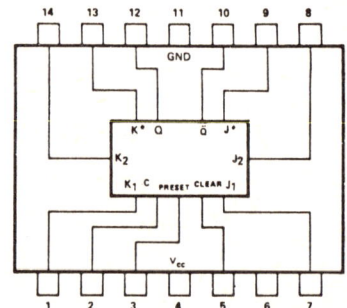

7472
J-K MASTER/SLAVE FLIP-FLOP

These J-K flip-flops are based on the master/slave principle. Each flip-flop has AND gate inputs for entry into the master section which is controlled by the clock pulse. The clock pulse also regulates the state of the coupling transistors which connect the master and slave sections. The sequence of operation is as follows:

1—Isolate slave from master.
2—Enter information from AND gate inputs to master.
3—Disable AND gate inputs.
4—Transfer information from master to slave.

TRUTH TABLE

t_n		t_{n+1}
J	K	Q
0	0	Q_n
0	1	0
1	0	1
1	1	$\bar{Q}_n$

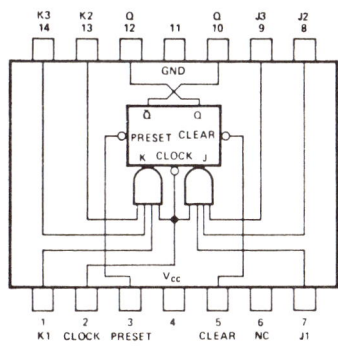

t_n = bit time before clock pulse
t_{n+1} = bit time after clock pulse

Total power dissipation: 50 mW
Operating frequency: 20 MHz

7473
DUAL J-K FLIP-FLOP

This negative-edge-clocked dual J-K flip-flop operates on the master/slave principle (see 7472). The device is quite useful for simple registers and counters where multiple J and K inputs are not required.

Input loading factor for clock and reset lines:	2
Total power dissipation:	80 mW
Propagation delay time:	30 nsec
Operating frequency:	15 MHz

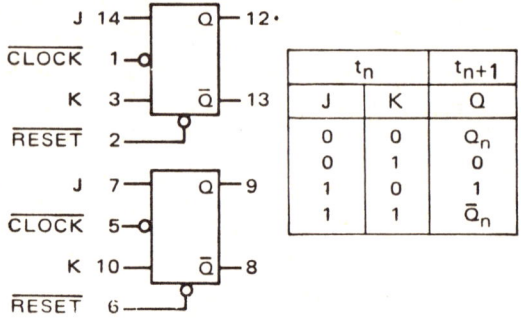

	t_n		t_{n+1}
	J	K	Q
	0	0	Q_n
	0	1	0
	1	0	1
	1	1	$\bar{Q}_n$

7474
DUAL D-TYPE EDGE-TRIGGERED FLIP-FLOP

This is a dual D-type edge-triggered flip-flop, featuring direct clear and preset inputs and complementary Q and $\bar{Q}$ outputs. Input information is transferred to the Q output on the positive-going edge of the clock pulse.

Clock triggering occurs at a voltage level of the clock pulse and is not directly related to the transition time of the positive-going pulse. After the clock input threshold voltage has been passed, the data input (D) is locked out.

A low input to the preset terminal sets Q to logic 1. A low input to the clear input sets Q to logic 0. Preset and clear are independent of the clock input.

Total power dissipation:	86 mW
Operating frequency:	25 MHz

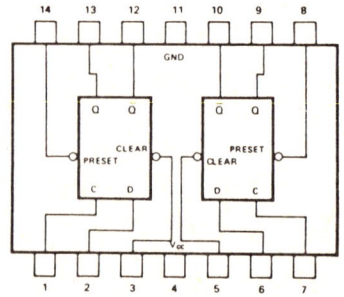

7475
QUAD LATCH

This device consists of four bistable latch circuits in one 16-pin package. Both Q and Q̄ outputs are available on all four devices. When the strobe (enable) is in the logic 1 state, the Q output will follow the state of the data input. When the strobe goes to the logic 0 state, the Q output will retain the state of the data input at the time of the transition from the logic 1 state.

Input loading factor for D: 2
Input loading factor for strobe: 4
Total power dissipation: 160 mW
Propagation delay time: 30 ns

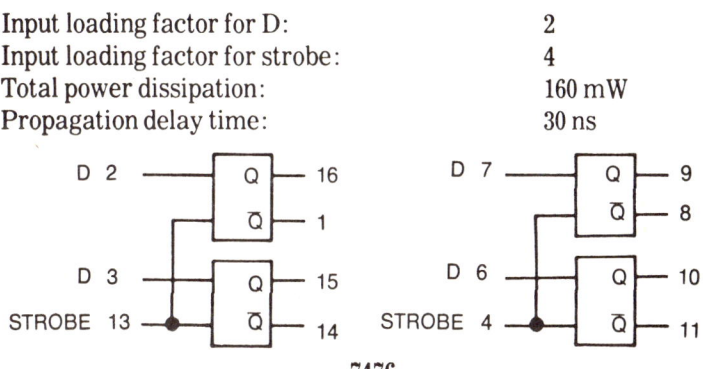

7476
DUAL J-K MASTER-SLAVE FLIP-FLOP

This J-K flip-flop is based on the master/slave principle (see 7472). Inputs to the master section are controlled by the clock pulse. The clock pulse also regulates the state of the coupling transistors which connect the master and slave sections.

A low input to the preset terminal sets Q to logic 1. A low input to the clear terminal sets Q to logic 0. Clear and preset are independent of the clock input.

Total power dissipation: 100 mW
Operating frequency: 20 MHz

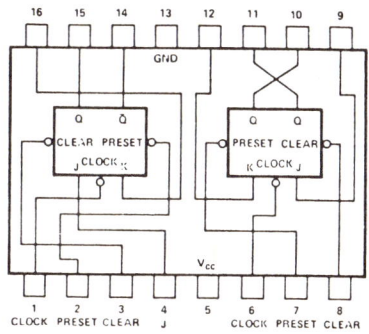

7486

QUAD 2-INPUT EXCLUSIVE OR GATE

The 7486 quad 2-input exclusive OR gate is a TTL element providing the function $A \oplus B = AB + \overline{AB}$ at the output.

Inputs		Output
A	B	Y
0	0	0
0	1	1
1	0	1
1	1	0

Total power dissipation: 150 mW
Propagation delay time: 18 nsec

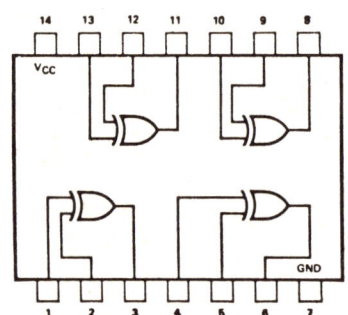

7489

64-BIT READ/WRITE MEMORY (RAM)

The 7489 is a TTL 64-bit read/write random-access memory, organized as 16 words of 4 bits each. The 7489 is ideally suited for application as scratch pads and high-speed buffer memories.

Words are selected through a 4-input binary decoder when the chip select input (CE) is at logic 0. Data is written into the memory when *read enable* (RE) is at logic 0, and read from the memory when RE is at logic 1.

Total power dissipation: 380 mW
Read Time: 35 nsec
Write Time: 50 nsec

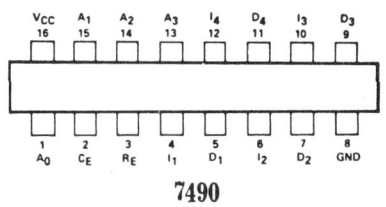

7490

DECADE COUNTER

The 7490 is a high-speed decade counter, consisting of four master/slave flip-flops internally connected to provide a divide-by-2 counter and a divide-by-5 counter. Gated direct-reset lines are provided to inhibit count inputs and return all outputs to a logic 0 or to a binary-coded decimal (BCD) count of 9. As the output from flip-flop A is not internally connected to the succeeding stages, the unit may be operated in three independent count modes:

1—When used as a BCD decade counter, the BD input must be externally connected to the A output. The A input receives the incoming count.

2—If a symmetrical divide-by-10 count is desired for frequency synthesizers or other applications requiring division of a binary count by a power of 10, the D output must be externally connected to the A input. The input count is then applied at the BD input, and a divide-by-10 square wave is obtained at output A.

3—For operation as a divide-by-2 counter and divide-by-5 counter, no external interconnections are required. Flip-flop A is used as a binary element for the divide-by-2 function. The BD input is used to obtain binary divide-by-5 operation at the B, C, and D outputs. In this mode, the two counters operate independently; however, all four flip-flops are reset simultaneously.

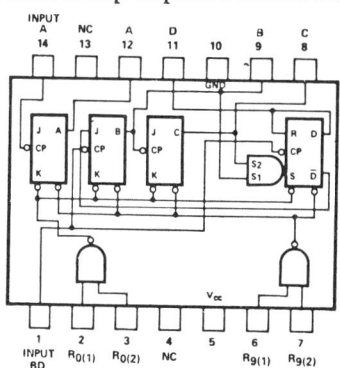

Total power dissipation:	160 mW
Operating frequency:	32 MHz

7492
DIVIDE-BY-TWELVE COUNTER

The 7492 is a high-speed 4-bit binary counter consisting of four master/slave flip-flops which are internally connected to provide a divide-by-2 counter and a divide-by-6 counter. A gated direct reset line is provided which inhibits the count inputs and simultaneously returns the four flip-flop outputs to a logic 0. As the output from flip-flop A is not internally connected to the succeeding flip-flops, the counter may be operated in two independent modes:

1—When used as a divide-by-12 counter, output A must be externally connected to input BD. The input count pulses are applied to input A. Simultaneous division of 2, 6, and 12 are performed at the A, C, and D outputs.

2—When used as a divide-by-6 counter, the input count pulses are applied to input BC. Simultaneously, frequency division of 3 and 6 are available at the C and D outputs. Independent use of flip-flop A is available if the reset function coincides with reset of the divide-by-6 counter.

Total power dissipation:	160 mW
Operating frequency:	32 MHz

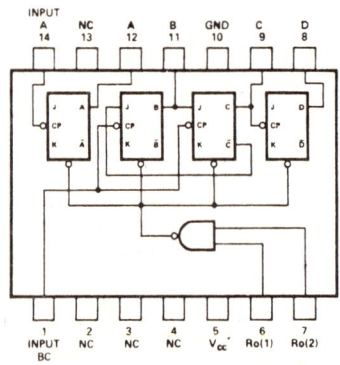

7493
4-BIT BINARY COUNTER

The 7493 is a high-speed 4-bit binary counter consisting of four master/slave flip-flops which are internally connected to provide a divide-by-2 counter and a divide-by-8 counter. A gated direct reset line is provided which inhibits the count inputs and simultaneously returns the four flip-flop outputs to a logic 0. As the output from flip-flop A is not internally connected to the succeeding flip-flops, the counter may be operated in two independent modes:

1—When used as a 4-bit ripple-through counter, output A must be externally connected to input B. The input count pulses are applied to input A. Simultaneous divisions of 2, 4, 8, and 16 are performed at the $A, B, C,$ and D outputs.

2—When used as a 3-bit ripple-through counter, the input count pulses are applied to input B. Simultaneous frequency divisions of 2, 4, and 8 are available at the $B, C,$ and D outputs. Independent use of flip-flop A is available if the reset function coincides with reset of the 3-bit ripple-through counter.

Total power dissipation: 160 mW
Operating frequency: 32 MHz

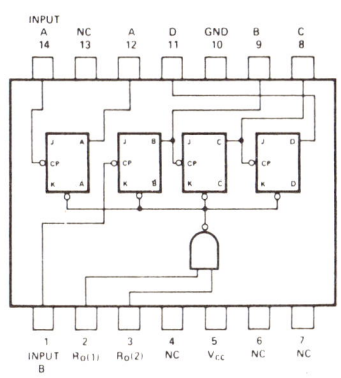

7495
4-BIT RIGHT/LEFT SHIFT REGISTER

The 7495 is a universal 4-bit shift register designed with standard TTL techniques. The circuit layout consists of 4 R-S master/slave flip-flops, 4 AND-OR-invert gates, and 6 inverters configured to form a versatile register which will perform right-shift/left-shift or parallel-in/parallel-out operations. These operations are controlled by input level to the mode control.

Right-shift operations are performed when a logic-0 level is applied to the mode control. Serial data is entered at the serial input (pin 1) and shifted one position right on each clock-1 pulse. In this mode, clock-2 and parallel inputs *A* through *D* are inhibited.

Parallel-in/parallel-out operations are performed when a logic-1 level is applied to the mode control. Parallel data is entered at parallel inputs *A* through *D* and is transferred to the data outputs *A* through *D* on each clock-2 pulse. In this mode, left-shift operations may be implemented by externally tying the output of each flip-flop to the parallel input of the previous flip-flop; with serial inputs, data is entered at input *D*.

Information must be present at the inputs prior to clocking, and transfer of data occurs on the falling edge of the clock pulse.

Total power dissipation:	195 mW
Operating frequency:	25 MHz

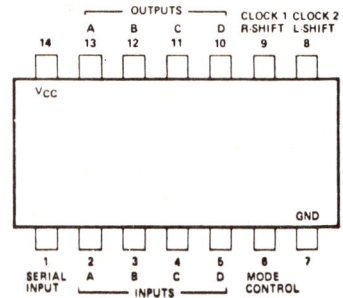

7496
5-BIT SHIFT REGISTER

This shift register consists of five R-S master/slave flip-flops connected to perform parallel-to-serial-to-parallel conversion of binary data. Since both inputs and outputs to all flip-flops are accessible, parallel-in/parallel-out or serial-in/serial-out operation may be performed.

All flip-flops are simultaneously set to the logic-0 state by applying a logic-0 voltage to the clear input. This condition may be applied independent of the state of the clock input.

The flip-flops may be independently set to the logic-1 state by applying a logic-1 to both the preset input of the specific flip-flop and the common preset input. The common preset input is provided to allow flexibility of either setting each flip-flop

206

independently or setting two or more flip-flops simultaneously. Preset is also independent of the state of the clock input or clear input.

Transfer of information to the output pins occurs when the clock input goes from a logic 0 to a logic 1. Since the flip-flops are R-S master/slave circuits, the proper information must appear at the R-S inputs of each flip-flop prior to the rising edge of the clock input voltage waveform. The serial input provides this information to the first flip-flop, while the outputs of the subsequent flip-flops provide information for the remaining R-S inputs. The clear input must be at a logic 1 and the preset input must be at a logic 0 when clocking occurs.

Total power dissipation: 240 mW
Operating frequency: 10 MHz

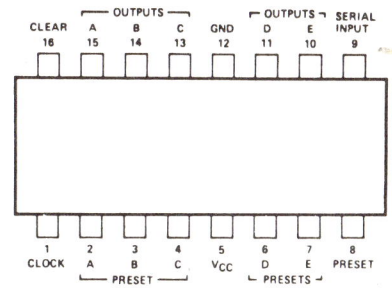

74107
DUAL J-K MASTER/SLAVE FLIP-FLOP

The 74107A J-K flip-flop is based on the master/slave principle (see 7472). Inputs to the master section are controlled by the clock pulse. The clock pulse also regulates the state of the coupling transistors which connect the master and slave sections.

Total power dissipation: 100 mW
Operating frequency: 20 MHz

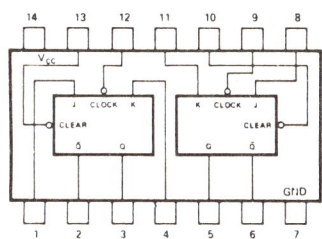

74141
BCD-TO-DECIMAL DECODER/DRIVER

This is a BCD-to-decimal decoder designed specifically to drive cold-cathode indicator tubes. This decoder demonstrates an improved capability to minimize switching transients in order to maintain a stable display.

Full decoding is provided for all possible input states. For binary inputs 10 through 15, all the outputs are off. Therefore the 74141, combined with a minimum of external circuitry, can use these invalid codes in blanking either leading or trailing zeros in a display.

Low-impedance diodes are also provided for each input to clamp negative-voltage transitions in order to minimize transmission-line effects.

Total power dissipation:	80 mW
Maximum open-collector output voltage:	60V
Maximum open-collector output current:	7 mA

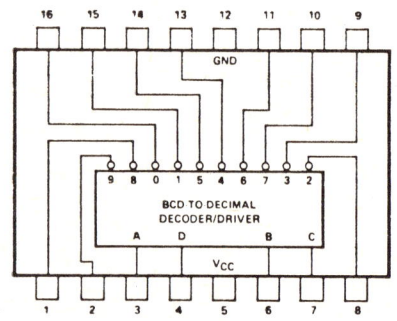

74150
16-LINE-TO-1-LINE
DATA SELECTOR/MULTIPLEXER

The 74150 is a 1-of-16 data selector which performs parallel-to-serial data conversion. The unit incorporates an enable circuit for chip select. This allows multiplexing from n lines to one line.

The 74150 is provided with a strobe input which when set to logic 0 enables the function of these multiplexers.

This data selector/multiplexer is fully compatible for use with other TTL or DTL circuits. A fan-out to 20 normalized series 74 loads is provided in the logic-1 state to facilitate connection of unused inputs to used inputs.

Total power dissipation: 200 mW
Propagation delay time (select) time: 22 nsec

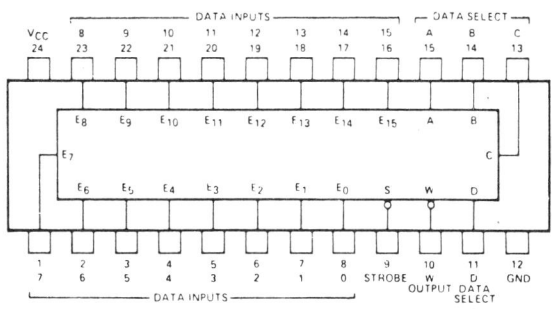

74151
8-LINE-TO-1-LINE
DATA SELECTOR/MULTIPLEXER

The 74151 is a 1-of-8 data selector which performs parallel-to-serial data conversion. The unit incorporates an enable circuit for chip select. This allows multiplexing from n lines to one line. Both true and complement outputs are available.

Total power dissipation: 145 mW
Propagation delay (select) time: 19 nsec

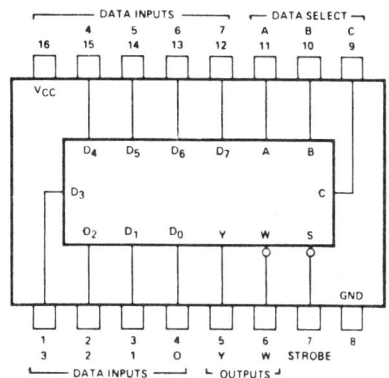

74153
DUAL 4-LINE-TO-1-LINE
DATA SELECTOR/MULTIPLEXER

Each of these monolithic data selector/multiplexers contains inverters and drivers to supply fully complementary on-chip binary-decoding and data selection. Separate strobe inputs are provided for each of the two 4-line sections.

These data selector/multiplexers are fully compatible for use with most TTL and DTL circuits. A fan-out to 20 series 74 loads is provided in the high-level state to facilitate connection of unused inputs to used inputs.

Total power dissipation: 180 mW
Propagation delay (select) time: 17 nsec

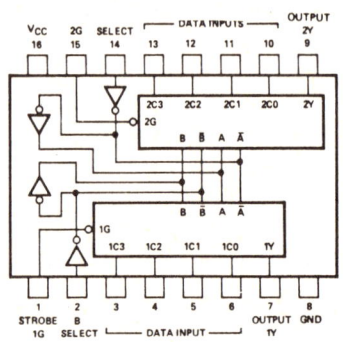

74155, 74156
DUAL 2-LINE-TO-4-LINE
DECODER/DEMULTIPLEXER

These monolithic TTL circuits feature dual 1-line-to-4-line demultiplexers with individual strobes and common binary address inputs, all in a single 16-pin package. When both sections are enabled by the strobes, the common binary address inputs sequentially select and route associated input data to the appropriate output of each section. The individual strobes permit activating or inhibiting each of the 4-bit sections as desired. The inverter following the 1C data input permits use as a 3-to-8-line decoder or 1-to-8-line demultiplexer without external gating.

The 74155 circuits are rated to fan out to 10 normalized series 74 loads in the low-level output state and to 20 loads in the high-level output state. The 74156 circuits, with open-collector outputs, are rated to sink 16 mA at a low-level output voltage of less than 0.4V. Input clamping diodes are provided on all of these circuits to minimize transmission-line effects and simplify system design.

Total power dissipation: 125 mW

Propagation delay time: 16 nsec (74155)

21 nsec (74156)

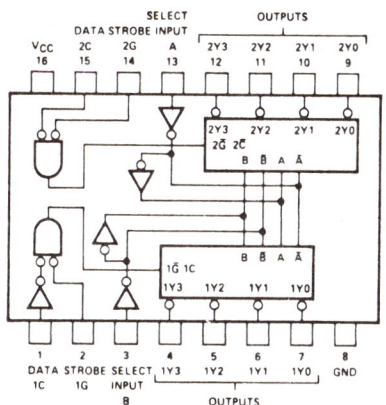

74157, 74158
QUADRUPLE 2-INPUT
DATA SELECTOR/MULTIPLEXER

The 74157 and 74158 are identical, with the exception of the outputs of the 74158 being inverted. These devices are logical implementations of a 4-pole/2-position switch, with the position of the switch being set by the logic levels supplied to the one select input. When the enable input (E) is high, all outputs are low, regardless of the other inputs.

The devices provide the ability, in one package, to select four bits of either data or control from two sources. By proper manipulation of the inputs, it can generate four functions of two variables, with one variable common. Thus any number can be replaced.

INPUTS				OUTPUT Y
ENABLE	SELECT	A	B	74157
H	X	X	X	L
L	L	L	X	L
L	L	H	X	H
L	H	X	L	L
L	H	X	H	H

H = High Level, L = Low Level, X = Irrelevant

211

Total power dissipation: 150 mW
Propagation delay (select) time: 14 nsec

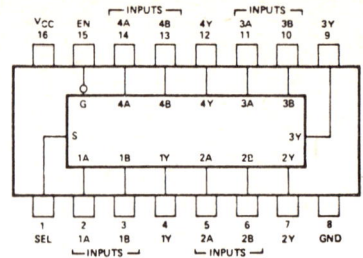

74160, 74161, 74162, 74163
SYNCHRONOUS 4-BIT COUNTER

These synchronous, presettable counters feature an internal carry look-ahead for application in high-speed counting schemes. The 74160 and 74162 are decade counters; the 74161 and 74163 are binary counters. Synchronous operation is provided by having all flip-flops clocked simultaneously, so that the output changes coincide with each other. This mode of operation eliminates the output counting spikes which are normally associated with asynchronous (ripple clock) counters. A buffered clock input triggers the four J-K master/slave flip-flops on the rising (positive-going) edge of the clock input waveform.

All inputs are diode-clamped to minimize transmission-line effects, thereby simplifying system design. A full fan-out to 10 normalized series 74 loads is available from each of the outputs in the low-level state. A fan-out of 20 normalized series 74 loads is provided in the high-level state to facilitate connection of unused inputs.

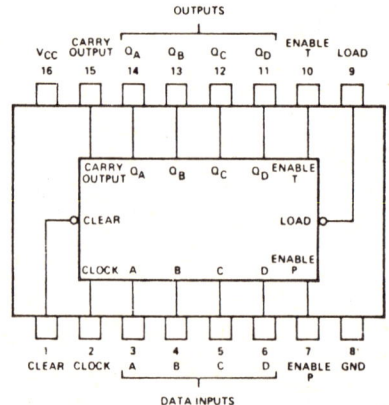

Total power dissipation:	305 mW
Operating frequency:	25 MHz

74164
8-BIT PARALLEL-OUT SERIAL
SHIFT REGISTERS

These 8-bit shift registers feature gated serial inputs and an asynchronous clear. The gated serial inputs (*A* and *B*) permit complete control over incoming data. A low at either or both of the inputs inhibits entry of new data and resets the first flip-flop to the low level at the next clock pulse. A high level on both inputs sets the first flip-flop to a high level at the next clock pulse. Data at the serial inputs may be changed while the clock is high. Information will be entered only on the low-to-high level transition of the clock input.

Total power dissipation:	167 mW
Maximum clock frequency:	25 MHz

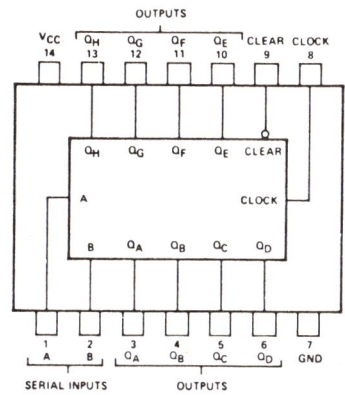

74165
PARALLEL-LOAD 8-BIT SHIFT REGISTER

The 74165 is an 8-bit serial shift register that shifts data to the right when clocked. Parallel-in access to each stage is made available by eight individual, direct data inputs which are enabled by a low level at the shift/load input. These registers also feature gated clock inputs and complementary outputs from the eighth bit.

Clocking is accomplished through a 2-input NOR gate, permitting one input to be used as a clock-inhibit function.

213

Holding either of the clock inputs high inhibits clocking, and holding either clock input low with the load input high enables the other clock input. The clock-inhibit input should be changed to the high level only while the clock input is high. Parallel loading is inhibited as long as the load input is high. When taken low, data at the parallel inputs is loaded directly into the register independently of the state of the clock.

Total power dissipation: 210 mW
Maximum clock frequency: 25 MHz

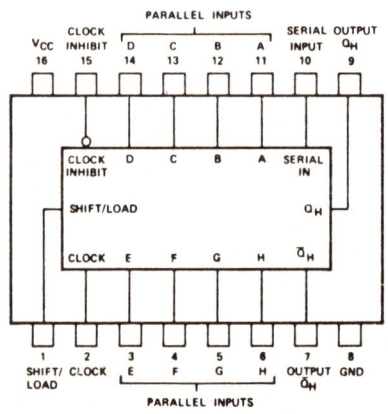

74180
8-BIT ODD/EVEN PARITY
GENERATOR/CHECKER

The 74180 8-bit odd/even parity generator and checker is a TTL monolithic array, featuring gating logic arranged to generate or check odd or even parity.

Total power dissipation: 170 mW
Propagation delay time: 35 nsec

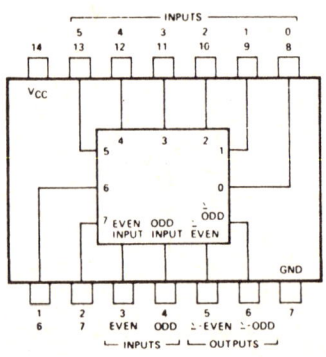

74192
SYNCHRONOUS DECADE UP/DOWN COUNTER

The outputs of the master/slave flip-flops are triggered by a low-to-high level transition of either count (clock) input. The direction of counting is determined by which count input is pulsed while the other count input is high.

These counters are fully programmable; that is, the outputs may be preset to any state by entering the desired data at the data inputs while the load input is low. The output will change to agree with the data inputs independently of the count pulses. This feature allows the counters to be used as modulo-n dividers by simply modifying the count length with the preset inputs.

A clear input has been provided which forces all outputs to the low level when a high level is applied. The clear function is independent of the count and load inputs. These counters were designed to be cascaded without the need for external circuitry. Borrow and carry outputs are available to cascade both the up and down counting functions. The borrow output produces a pulse equal in width to the countdown input when the counter underflows. Similarly, the carry output produces a pulse equal in width to the countup input when an overflow condition exists. The counters can then be easily cascaded by feeding the borrow and carry outputs to the countdown and countup inputs, respectively, of the succeeding counter.

Total power dissipation: 325 mW
Maximum count frequency: 32 MHz

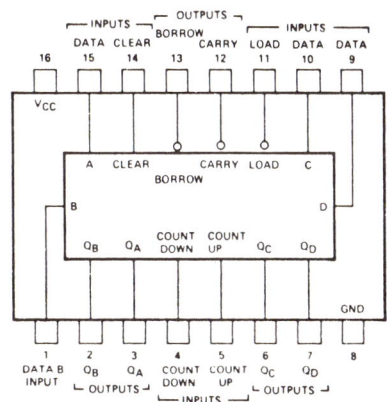

Chapter 8

Linear IC Types

This chapter treats linear ICs in much the same way the preceding chapter treated digital ICs. It explains their specifications and presents specific information on the most popular types.

Linear ICs (LICs) are used to perform a wide variety of functions. In general, they employ direct coupling, differential amplifier stages, and negative feedback—all of which are discussed in preceding chapters. General-purpose LICs such as operational amplifiers are extremely versatile devices that may be adapted to provide many different circuit functions. Special-purpose LICs are designed to perform specific circuit functions, usually taking the place of several stages of discrete components in a high-volume application such as the sound circuitry of a TV receiver. Special-purpose circuits are primarily of interest to manufacturers. We only consider general-purpose LICs here, since these are the ones most readers are apt to encounter.

There is no standard line of linear ICs, no LIC counterpart of the 54/74 digital IC series. There is not even a standard line of general-purpose LICs. However, the data given here covers a representative sampling of general-purpose LICs, including the most popular ones.

COMPENSATION

It has already been explained how the closed-loop frequency response of an op-amp can be shaped by components in the feedback circuit. The *open*-loop response can also be shaped, as shown by the curves in the data sheets at the end of this chapter. The open-loop response can be changed by inserting a series resistor/capacitor combination across the input, across an internal component such as a transistor (phase lag compensation), or between the input and output of an inverting gain stage in the IC (Miller-effect phase lag compensation). The open-loop response can also be altered by a capacitor connected between the collectors in one of the IC's high-gain stages (phase lead compensation). Comprehensive data sheets such as those given here show how external components are to be connected to the IC for compensation and show the frequency response for various recommended values of compensating components.

The purpose of phase compensation is to prevent *instability*; that is, a tendency of the circuit to operate erratically or break into oscillation. As the frequency of operation increases, the phase shift also increases. Eventually, a frequency is reached at which the phase shifts 180 degrees (positive feedback), the precondition for oscillation. Compensation limits the phase shift to prevent instability, but it does so at the expense of the frequency response.

For an understanding of the response curves, refer to Fig. 8-1. Assume that the ratio of feedback components has been chosen so that the open-loop gain is 60 dB. If the compensation components are chosen to give response curve 3, the frequency response of the amplifier will be flat out to about 40 kHz. Then the response will follow the rolloff of curve 3, dropping off at a

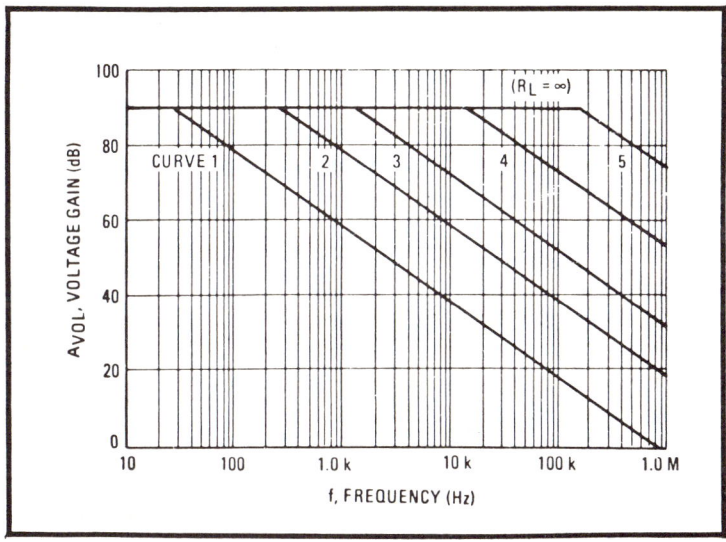

Fig. 8-1. Frequency response curves for Motorola HEP C6015L operational amplifier.

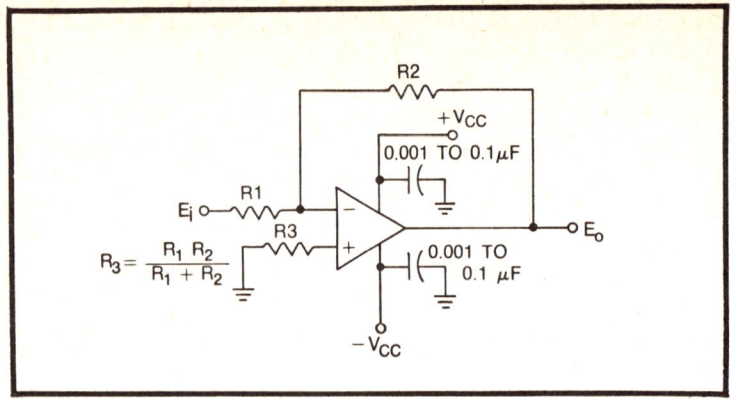

Fig. 8-2. Basic operational amplifier circuit.

rate of about 6 dB per octave (doubling of frequency), or 20 dB per decade.

In designing with operational amplifiers, it is important to remember that the required closed-loop gain should always be less than the open-loop gain at a given frequency. Thus, if the amplifier had the response shown by curve 2, the closed-loop gain at 100 kHz would have to be less than 40 dB, the approximate open-loop gain at that frequency.

SPECIFICATIONS AND CIRCUIT DESIGN

Circuit design with ICs is fairly simple if you have the specifications of the IC and know to use them. To illustrate this, consider the basic operational amplifier circuit of Fig. 8-2.

The two capacitors are for decoupling; that is, for bypassing signals around the power supply. These capacitors are used at the IC, as close to its power supply terminals as possible. They prevent the signals of a number of ICs operated from a common power supply from being coupled together by way of the supply. The proper values for them may be given on the data sheet. Generally they are ceramic disc capacitors of between 0.001 and 0.1 μF. If no value is given, try the larger value listed above.

The ratio of R_2 to R_1 determines the gain of the amplifier, as already explained. The maximum value of R_1 depends on a specification called the *input bias current*. And the value of R_3, in turn, depends on the values of R_1 and R_2. To determine the proper values for these resistors, we must refer to the specifications of the IC.

Maximum Supply Voltage

There is a maximum voltage specification that gives the maximum voltage that can appear across the supply terminals of the IC without damaging it. Most IC power supplies have a voltage output tolerance of 5% to 10%, and linear ICs will usually operate satisfactorily with a supply having a tolerance of 20%. It is important to consider the tolerance in determining whether a given power supply may be used with a certain IC. If a 10V power supply with a tolerance of 20% is used with an IC having a maximum supply voltage of 10V, the IC may be damaged when the power supply voltage reaches 12V. If the supply voltage were very much more than 10V. the IC would be damaged instantly.

Nominal Supply Voltage

This is the supply voltage at which an IC is tested and the one at which the "typical" specifications apply. At higher voltages (but less than the maximum), an IC will exceed its typical performance ratings; at lower voltages, it will underperform.

Total Device Dissipation

This is the maximum amount of power an IC can handle at a nominal ambient temperature (typically 25°C, or room temperature) without being damaged by overheating. Ordinarily, a *derating* figure of so many milliwatts per degree Celsius is given with this specification. For example, if the derating figure is 2 mW/°C, then the IC can handle 10 mW less power at 30°C than at 25°C. The amount of power an IC is actually handling is equal to the difference between its input power (power supply voltage times power supply current) and its output power (load voltage times load current). For example, if the input power to an op-amp is 100 mW at the time it is delivering 40 mW to a load, then the op-amp is dissipating 60 mW.

Temperature Range

Often two temperature-range specifications are given: one for storage and one for operation. The operating temperature range is typically 0°C to 75°C, and the storage temperature range is somewhat greater. High temperatures are more damaging than low temperatures; no IC should ever have a temperature in excess of 200°C. Operating an IC at excessive temperatures may cause premature failure or changes in the operating characteristics. In some cases, if the heating is not too severe, the IC may again operate normally when its temperature is lowered to the recommended range.

Maximum Input Signal

This and the other ratings given so far (except the nominal supply voltage) are maximum ratings that must not be exceeded if damage to the IC is to be avoided. The maximum input signal may be expressed as a peak voltage having equal positive and negative values; for example, $\pm 18V$. Alternatively, separate positive and negative values may be given; for example, $+2V$ and $-5V$.

Device Dissipation

Device dissipation is the amount of power consumed by the IC itself when in the quiescent, or unloaded, condition. The load power plus the dissipation of the IC itself must not exceed the total device dissipation. Therefore, the lower the device dissipation is for a given total device dissipation, the greater the amount of power that can be delivered to the load.

Input Bias Current

With a linear amplifier connected into a circuit, two currents flow into the input of the amplifier, as shown by Fig. 8-3. The average value of these currents, or $(I_1 + I_2)/2$, is termed the *input bias current*. This specification determines the value to be used for the smaller closed-loop resistor. For example, the value of R_1 in Fig. 8-2 should be chosen so that the voltage dropped across R_1 by the input bias current is much less than the input signal, usually less than 10% of it. Once R_1 is chosen, R_2 can be chosen to give the desired closed-loop gain, and R_3 is determined roughly by the formula in Fig. 8-2.

Input Offset Current

The *input offset current* is the difference between the two input bias currents: $I_1 - I_2$, or vice versa. If either of the bias currents is much larger than the other, the unequal voltages dropped across the input resistors then appear to the amplifier as a signal voltage which is thus amplified. Since the input bias currents vary with temperature, this input imbalance affects the temperature stability of the op-amp circuit. A resistor (R3 in Fig. 8-2) is connected between the noninverting input and ground and its value is selected to minimize this imbalance. An appropriate value may be estimated by the formula in Fig. 8-2.

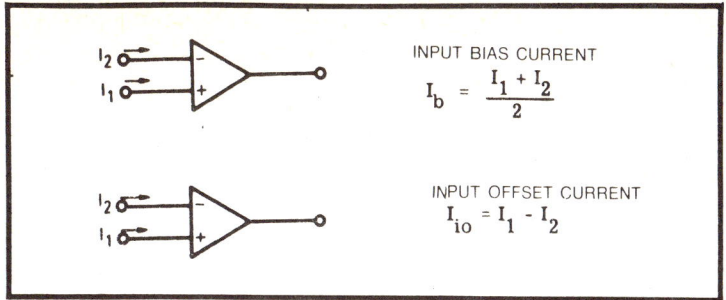

Fig. 8-3. Input bias current.

The value of R3 can also be varied to compensate for the input offset voltage.

Input Offset Voltage

As shown in Fig. 8-4, the input offset voltage, E_{io}, is equal to the DC voltage that must applied between the input terminals to set the output voltage equal to zero. The offset voltage is, in reality, a measure of the error in the differential input circuitry. If both inputs are grounded, the differential input voltage is zero and the output should also be zero. But if the offset voltage is 1 mV, a typical value, then the op-amp will amplify this offset voltage and the output of the op-amp will not be zero. If the op-amp has a gain of 1000, for example, then the output voltage of the op-amp will be 1000×1 mV = 1V. For lower op-amp gains, the effect of the offset voltage is less significant.

Some ICs have an offset adjustment terminal to which an external DC voltage is applied through a potentiometer to compensate for the offset voltage. In this case, the IC data sheet will specify the voltage, connection, and potentiometer value. In other cases, you can compensate for offset voltage effects by varying the value of R3 in Fig. 8-2; this introduces a voltage drop, due to the input bias current, which neutralizes the offset voltage.

Maximum Output Voltage Swing

This is the peak or peak-to-peak output that can be obtained without clipping. If the input voltage times the

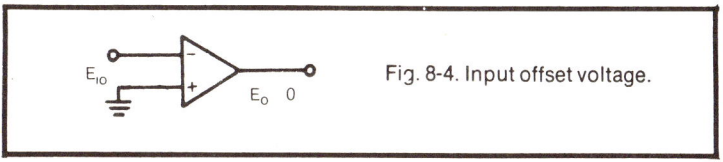

Fig. 8-4. Input offset voltage.

amplifier gain exceeds this rating, clipping (but not damage to the IC) will occur. This rating is related to *slew rate* and compensation. The slew rate is the maximum rate at which output voltage can change with respect to time without causing the IC to operate nonlinearly. The slew rate is another specification of the IC, but it can be changed by compensation measures. A large value of compensation capacitor will reduce the slew rate and hence the maximum output voltage swing.

Input Impedance

The *input impedance* is the impedance "looking into" one input of a linear IC with the other input grounded, generally measured at 25°C and 1 kHz. The input impedance of the IC should normally be much greater than the impedance of the source that is feeding it, otherwise the IC will load down the circuit it is connected to.

Output Impedance

The *output impedance* is the impedance seen by a load at the output of an IC amplifier. Specification sheets give the open-loop output impedance. The *closed*-loop output impedance, which should be as low as possible and always less than the load impedance, is given by

$$Z_{OUT(CL)} = \cfrac{Z_{OUT(OL)}}{1 + \cfrac{A_{OL}R_1}{R_1 + R_2}}$$

where $Z_{OUT(OL)}$ = open-loop output impedance
A_{OL} = open-loop voltage gain
R_1, R_2 = closed-loop resistors (Fig. 8-2)

FAIRCHILD μA702A OPERATIONAL AMPLIFIER

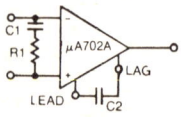

Phase compensation circuit.

Curve	R1	C1	C2
1	∞	0	50 pF
2	∞	0	50 pF
3	200Ω	1000 pF	50 pF
4	200Ω	1000 pF	50 pF

Phase compensation values.

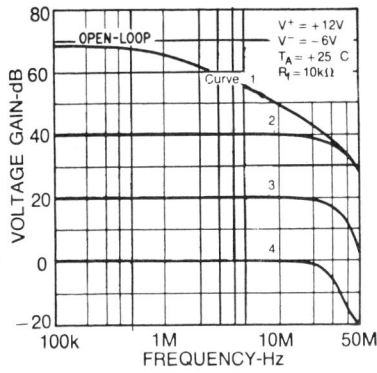

Frequency response for various compensation values.

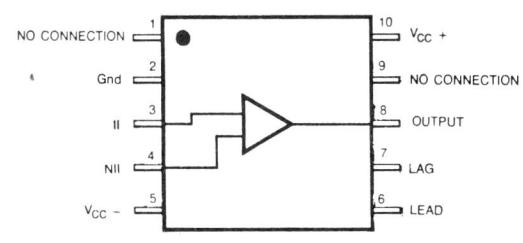

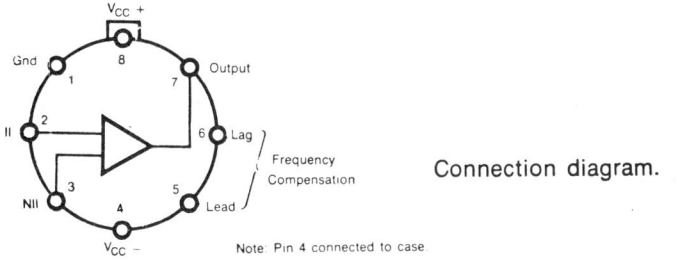

Connection diagram.

Note: Pin 4 connected to case.

Maximum supply voltage: 21V (V+ to V−)
Nominal supply voltage: +12V, −6V
Total device dissipation: 200 mW
Temperature range: −55°C to +125 C
Input offset voltage: 0.5 mV
Input offset current: 180 nA
Input bias current: 2 μA
Device dissipation: 90 mW
Open-loop voltage gain: 71 dB
Useful frequency range: 30 MHz
Common-mode rejection: 100 dB
Maximum output voltage swing: 11V
Input impedance: 40K
Output impedance: 200Ω
Maximum input signal: −4V, +5V

223

FAIRCHILD μA709 OPERATIONAL AMPLIFIER

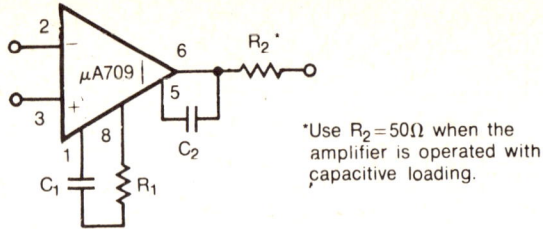

*Use $R_2 = 50\Omega$ when the amplifier is operated with capacitive loading.

Phase compensation circuit.

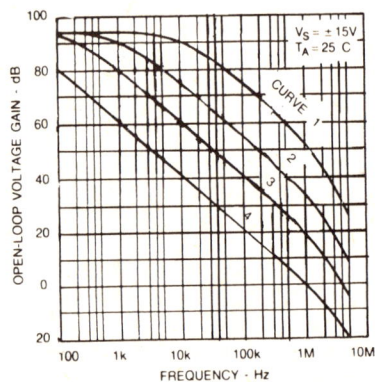

Frequency response for various compensation values.

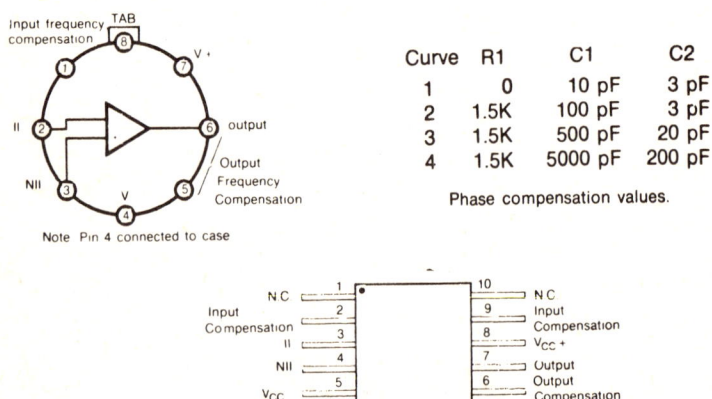

Note Pin 4 connected to case

Curve	R1	C1	C2
1	0	10 pF	3 pF
2	1.5K	100 pF	3 pF
3	1.5K	500 pF	20 pF
4	1.5K	5000 pF	200 pF

Phase compensation values.

Connection diagram.

Maximum supply voltage:	±18V
Nominal supply voltage:	±15V
Total device dissipation:	300 mW
Temperature range:	−55°C to +125°C
Input offset voltage:	1 mV
Input offset current:	50 nA
Input bias current:	200 nA
Device dissipation:	80 mW
Open-loop voltage gain:	93 dB
Common-mode rejection:	90 dB
Maximum output voltage swing:	28V p-p
Input impedance:	400K
Output impedance:	150Ω
Maximum input signal:	±10V

FIARCHILD μA739C OPERATIONAL AMPLIFIER

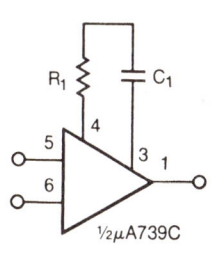

Phase compensation circuit.

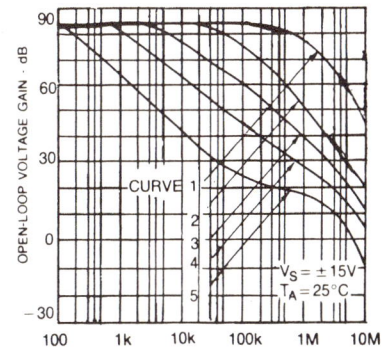

Frequency response for various compensation values.

Curve	R1	C1
1	∞	0
2	470Ω	300 pF
3	150Ω	1000 pF
4	33Ω	0.01 μF
5	4.7Ω	0.1 μF

Phase compensation values.

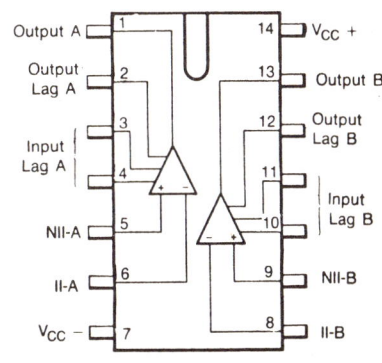

Connection diagram.

225

Maximum supply voltage:	±18V
Nominal supply voltage:	±15V
Total device dissipation:	500 mW
Temperature range:	0°C to 70°C
Input offset voltage:	1 mV
Input offset current:	50 nA
Input bias current:	30 nA
Device dissipation:	270 mW
Open-loop voltage gain:	86 dB
Common-mode rejection:	90 dB
Maximum output voltage swing:	+13V, −15V
Input impedance:	150K
Output impedance:	5K
Maximum input signal:	±15V

**FAIRCHILD μA741
OPERATIONAL AMPLIFIER
(INTERNALLY COMPENSATED)**

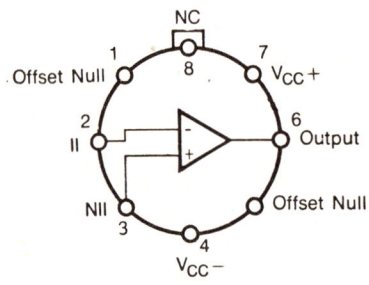

NOTE: PIN 4 CONNECTED TO CASE

Connection diagram.

Maximum supply voltage:	±22V
Nominal supply voltage:	±15V
Total device dissipation:	500 mW
Temperature range:	−55°C to +125°C
Input offset voltage:	1 mV
Input offset current:	30 nA
Input bias current:	200 nA
Device dissipation:	50 mW
Open-loop voltage gain:	106 dB
Useful frequency range:	1 MHz
Common-mode rejection:	90 dB
Maximum output voltage swing:	28V p-p
Input impedance:	1M
Maximum input signal:	±15V

FAIRCHILD µA747
OPERATIONAL AMPLIFIER

(Dual 741)

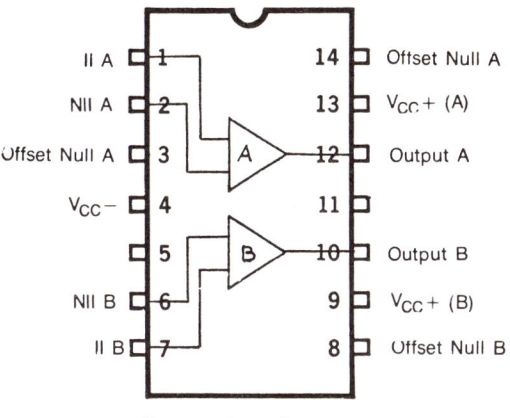

Connection diagram.

Maximum supply voltage:	±22V
Nominal supply voltage:	±15V
Total device dissipation:	500 mW
Temperature range:	−55°C to +125°C
Input offset voltage:	1 mV
Input offset current:	30 nA
Input bias current:	200 nA
Device dissipation:	50 mW
Open-loop voltage gain:	106 dB
Common-mode rejection:	90 dB
Maximum output voltage swing:	28 V p-p
Input impedance:	1M
Maximum input signal:	+7V, −12V

MOTOROLA HEP C6015L
DUAL OPERATIONAL AMPLIFIERS

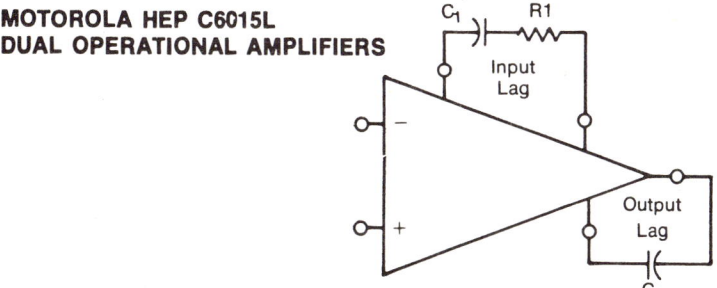

Phase compensation circuit.

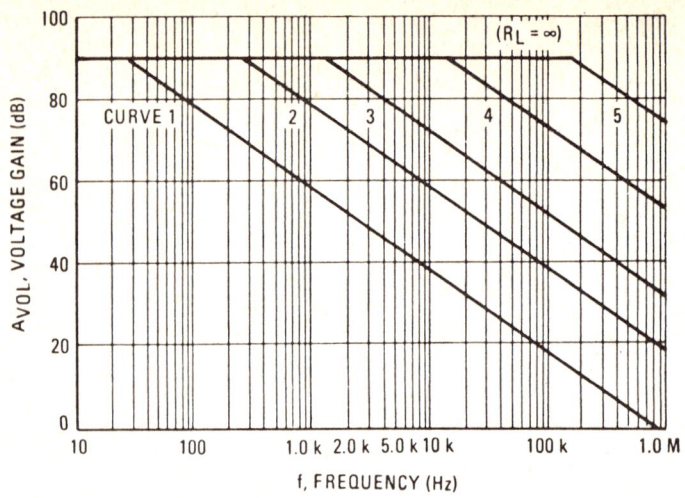

Frequency response for various compensation values.

Curve	R1	C1	C2
1	1.5K	5000 pF	200 pF
2	1.5K	500 pF	20 pF
3	1.5K	100 pF	3 pF
4	0	10 pF	3 pF
5	∞	0	3 pF

Phase compensation values.

Connection diagram.

Maximum supply voltage:	±18V
Nominal supply voltage:	±15V
Total device dissipation:	750 mW
Temperature range:	0°C to 75°C
Input offset voltage:	1.0 mV
Input offset current:	50 nA
Input bias current:	400 nA
Device dissipation:	160 mW
Open-loop voltage gain:	93 dB
Useful frequency range:	50 MHz
Common-mode rejection:	100 dB
Maximum output voltage swing:	28V p-p
Input impedance:	150K
Output impedance:	30Ω
Maximum input signal:	±18V

228

MOTOROLA HEP C6052G
OPERATIONAL AMPLIFIER
(Internally Compensated 741)

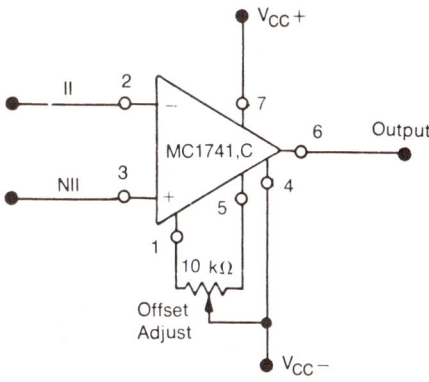

Connection diagram for TO package.

Maximum supply voltage:	±18V
Nominal supply voltage:	±15V
Total device dissipation:	680 mW (TO can)
Temperature range:	−55 to +125V
Input offset voltage:	2.0 mV
Input offset current:	30 nA
Input bias current:	200 nA
Device dissipation:	50 mW
Open-loop voltage gain:	100 dB
Useful frequency range:	1 MHz
Common-mode rejection:	90 dB
Maximum output voltage swing:	28V p-p
Input impedance:	1M
Output impedance:	75Ω
Maximum input signal:	±15V

MOTOROLA HEP C6053 OPERATIONAL AMPLIFIER

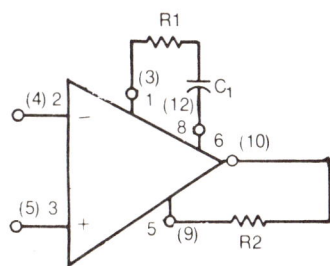

Phase compensation circuit.

229

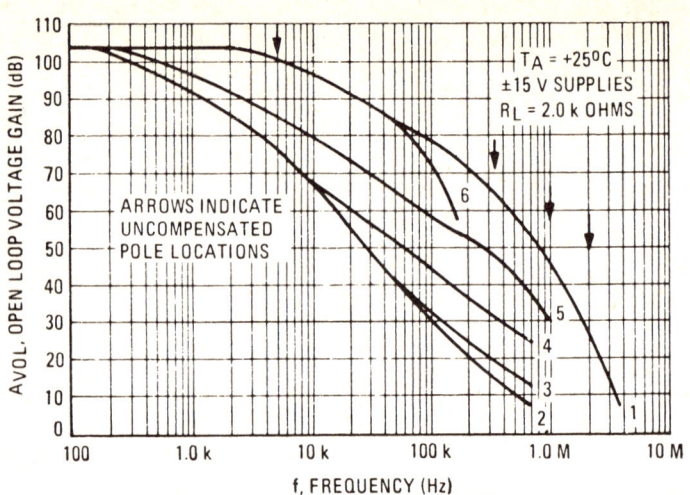

Frequency response for various compensation values.

Curve	R1	R2	C1
1	∞	10K	0
2	390Ω	10K	2200 pF
3	1K	10K	2200 pF
4	10K	10K	2200 pF
5	30K	10K	1000 pF
6	0	10K	10 pF

Phase compensation values.

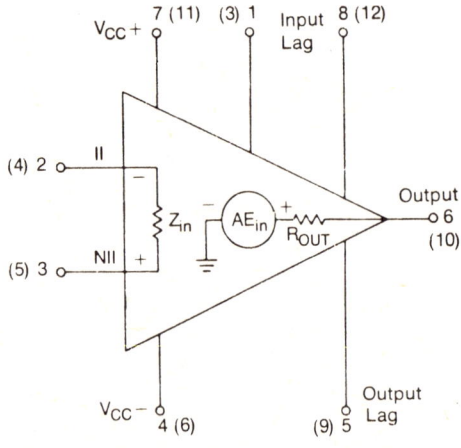

Connection diagram. (Pin numbers in parentheses refer to DIP; others refer to TO package.)

Maximum supply voltage:	±18V
Nominal supply voltage:	±15V
Total device dissipation:	680 mW (TO can)
Temperature range:	575 mW (DIP)
Input offset voltage:	0°C
Input offset current:	2 mV
Input bias current:	20 nA
Device dissipation:	200 nA
Open-loop voltage gain:	90 mW
Useful frequency range:	100 dB
Common-mode rejection:	5 MHz
Maximum output voltage swing:	110 dB
Input impedance:	300 K
Output impedance:	4K
Maximum input signal:	±18V

RCA CA3010A OPERATIONAL AMPLIFIER

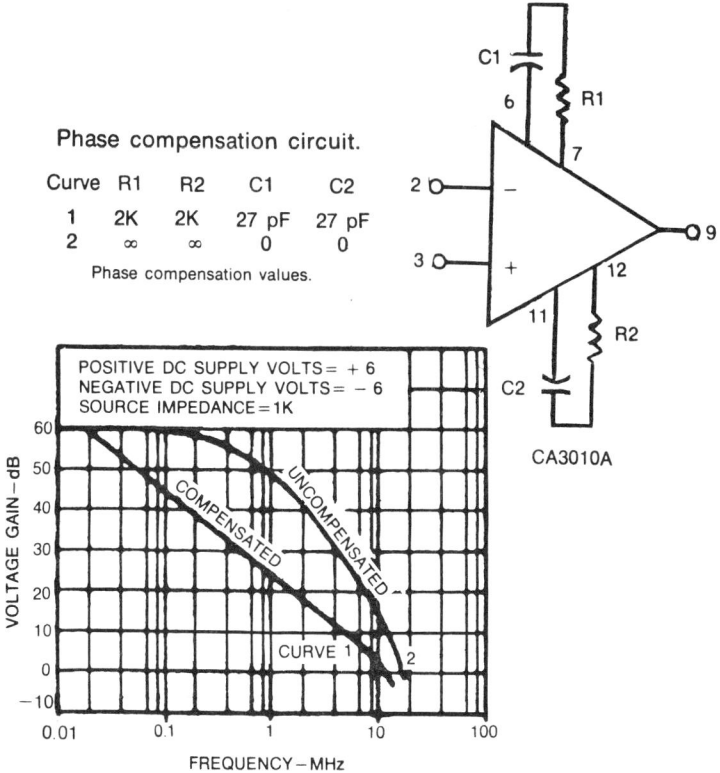

Phase compensation circuit.

Curve	R1	R2	C1	C2
1	2K	2K	27 pF	27 pF
2	∞	∞	0	0

Phase compensation values.

POSITIVE DC SUPPLY VOLTS = + 6
NEGATIVE DC SUPPLY VOLTS = − 6
SOURCE IMPEDANCE = 1K

CA3010A

Frequency response for various
compensation values.

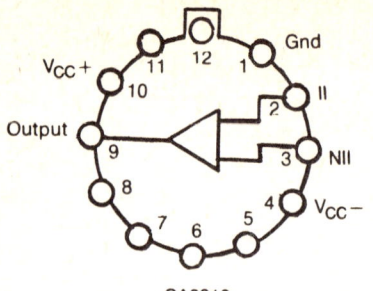

CA3010

Connection diagram.

Maximum supply voltage:	±10V
Nominal supply voltage:	±6V
Total device dissipation:	300 mW
Temperature range:	+55°C to +125°C
Input offset voltage:	0.9 mV
Input offset current:	0.3 μA
Input bias current:	2.5 μA
Device dissipation:	30 mW
Open-loop voltage gain:	60 dB
Useful frequency range:	15 mHz
Common-mode rejection:	94 dB
Maximum output voltage swing:	6.75V p-p
Input impedance:	20K
Output impedance:	160Ω
Maximum input signal:	+0.5V, −4V

RCA CA3015A

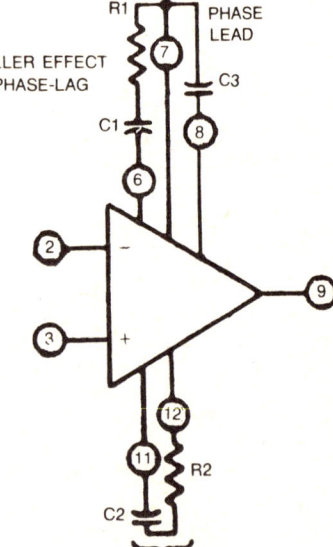

Phase compensation circuit.

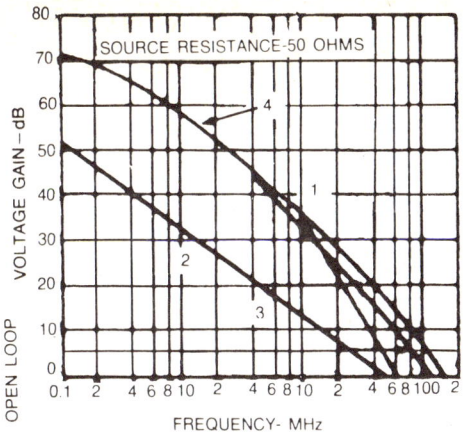

Frequency response for various
compensation values.

Curve	R1	R2	C1	C2	C3
1	∞	∞	0	0	0
2	820	820	18 pF	18 pF	0
3	∞	∞	0	0	10 pF
4	∞	∞	0	0	47-1000 pF

Phase compensation values.

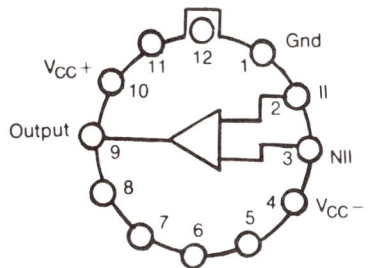

Connection diagram.

Maximum supply voltage:	±20V
Nominal supply voltage:	±12V
Total device dissipation:	600 mW
Temperature range:	−55 to +125°C
Input offset voltage:	1 mV
Input offset current:	0.5 μA
Input bias current:	4.7 μA
Device dissipation:	175 mW
Open-loop voltage gain:	70,2 dB
Useful frequency range:	50 MHz

Common-mode rejection: 103.5 dB
Maximum output voltage swing: 14V p-p
Input impedance: 10K
Output impedance: 85Ω
Maximum input signal: +0.65V, −8V

RCA CA3029A CA3037A OPERATIONAL AMPLIFIER

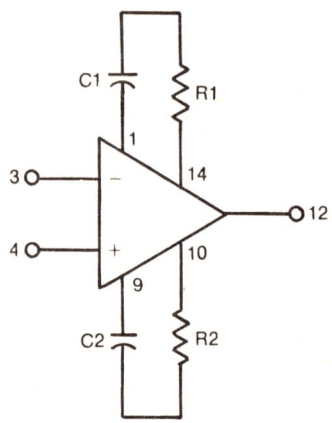

Phase compensation circuit.

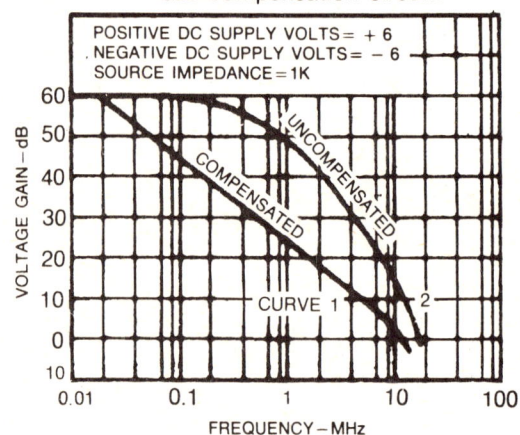

Frequency response for various
compensation values.

Curve	R1	R2	C1	C2
1	2K	2K	27 pF	27 pF
2	∞	∞	0	0

Phase compensation values.

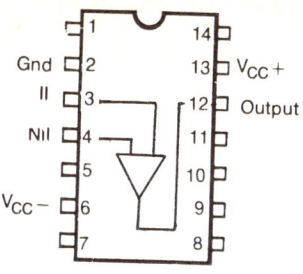

Connection diagram.

Maximum supply voltage: ± 6V
Nominal supply voltage: 300 mW
Total device dissipation: 0 to 70°C (CA3029A)
Temperature range: − 55 to + 125°C (CA3037A)
Input offset voltage: 0.9 mV
Input offset current: 0.3 μA
Input bias current: 2.5 μA
Device dissipation: 30 mW

Open-loop voltage gain: 60 dBV
Useful frequency range: MHz
Common-mode rejection: 94 dB
Maximum output voltage swing: 6.75V
Input impedance: 20K
Output impedance: 160Ω
Maximum input signal: + 0.5V, − 4V

RCA CA3030A, CA3038A   OPERATIONAL AMPLIFIER

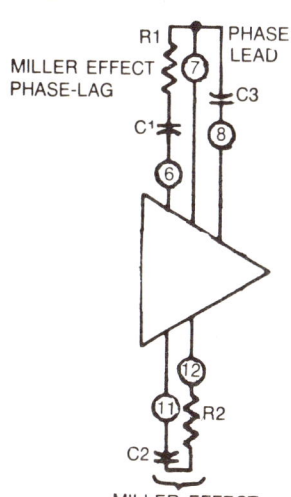

Phase compensation circuit.

235

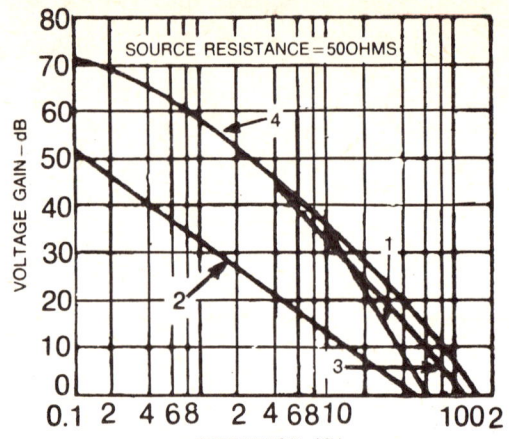

Frequency response for various
compensation values.

Curve	R1	R2	C1	C2	C3
1	∞	∞	0	0	0
2	820	820	18 pF	18 pF	0
3	∞	∞	0	0	10 pF
4	∞	∞	0	0	47-1000 pF

Phase compensation values.

Connection diagram.

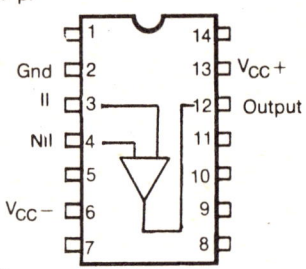

Maximum supply voltage:	±20V
Nominal supply voltage:	±12V
Total device dissipation:	600 mW
Temperature range:	0 to 70°C (CA3030A)
	−55 to +125°C (CA3038A)
Input offset voltage:	1 mV
Input offset current:	0.5 μA
Input bias current:	4.7 μA
Device dissipation:	175 mW
Open-loop voltage gain:	702 dB
Useful frequency range:	50 MHz
Common-mode rejection:	103.5 dB
Maximum output voltage swing:	14V p-p
Input impedance:	10K
Output impedance:	85Ω
Maximum input signal:	+0.65V, −8V

Chapter 9

IC Projects and Experiments

The best way to understand how integrated circuits work is by actually using them. For this reason, a number of experiments and projects have been included in this chapter that stress the educational side of ICs. For example, the first project is the construction of a digital logic demonstrator that will help you to understand how the various logic gates work. Other projects will enable you to learn more about op-amps, oscillators, transmitters and receivers, audio and video amplifiers, multivibrators, etc.

Every attempt has been made to select projects that use components which are easily obtained in electronics hobby stores such as Radio Shack and Lafayette Radio. Most of the integrated circuits are produced by several manufacturers, so you should have no trouble in making substitutions. Although one of the projects is the construction of a power supply, all you really need are a few flashlight batteries to make the circuits work.

DIGITAL LOGIC DEMONSTRATOR

A quick way to learn about the basic logic functions is to build the circuit shown in Figs. 9-1 and 9-2. This circuit illustrates some of the principles discussed in Chapters 3 and 4. If you are new to electronics experimenting, the pictorial/schematic drawings will help you to get started. The circuit can be quickly built on a piece of perforated circuit board measuring 8 by 5 inches. For convenience, solderless spring terminals can be used to make connections using pieces of wire or clip leads.

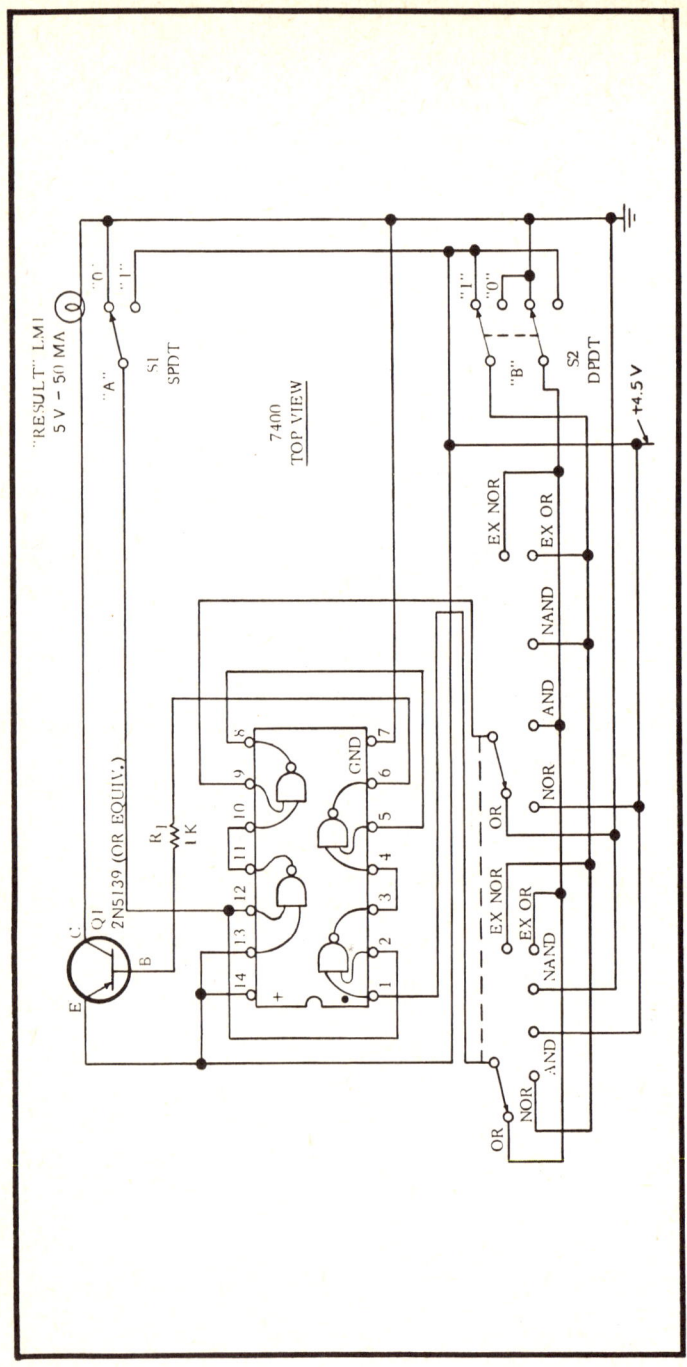

Fig. 9-1. Digital logic demonstrator—circuit layout. (Courtesy Grantham School of Engineering)

The pictorial diagram in Fig. 9-2 shows the placement of the components. A length of bus wire or a metal strap running across the top and bottom of the perforated board is very useful for making connections to the positive and negative terminals of the power supply or batteries. A 5V power supply is desirable, but three 1.5V flashlight batteries connected in series will do just as well.

In constructing this circuit, I used sockets for the lamp and integrated circuit. This approach helps to prevent damage to these components since they would otherwise be difficult to solder to. You might consider this to be a luxury, however, and choose to omit them. I also used regular toggle switches for the two double-pole/double-throw (DPDT) switches in the circuit, but rocker switches or even clip leads could also be used.

Note that the top view of the type 7400 IC is illustrated in the two figures. Be especially careful when connecting battery voltage that you connect the right polarity to the right terminals, for this is the only time that it is easy to damage the IC. After the power has been connected, any other pin of the IC can be connected to either the positive or negative terminal of the battery and no damage will result to the IC. Even the output terminals (pins 3, 6, 8, and 11) may be connected to either side of the battery, although this will make the IC heat up a little; it is not recommended, but no damage will occur if you should do it accidentally.

The various logic circuits are then made by connecting a jumper wire or clip lead as illustrated in Fig. 9-2. To make an OR circuit, for example, a clip lead is connected from pin 1 of the IC to the left-hand OR terminal on the board and the second clip lead is connected from pin 9 to the right-hand OR terminal. The clip leads connected to pins 1 and 9 of the IC will remain there; the other ends of the clips are moved to the right, one position at a time, to demonstrate the OR, NOR, AND, NAND, exclusive OR, and exclusive NOR. It would be a good idea to label these connections clearly on the perforated board. When the circuit has been constructed, proceed to the following experiments.

The OR Gate

Make sure that both switches of the logic demonstrator are in the 0 or *off* position and that the clip leads are connected to the OR terminals, as explained above. Apply power to the circuit. Now place switch *A* or *B* to the 1 position—lamp *L* should light. If the lamp does not light, check the connections of your circuit and see that they conform to the schematic.

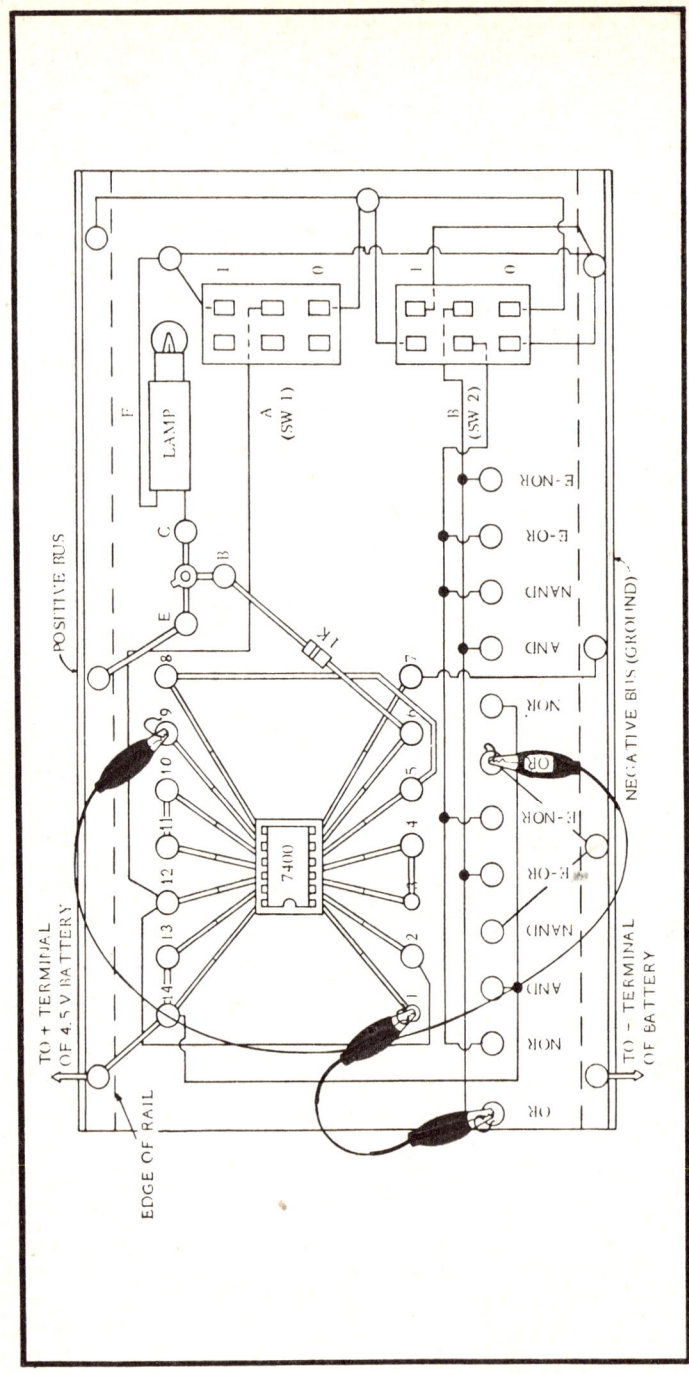

Fig. 9-2. Digital logic demonstrator—component layout showing relative positions of 7400 IC, lamp, switches, and spring connectors. (Courtesy Grantham School of Engineering)

Switch *A* and *B* to the various states shown in the table below and record the condition of lamp *L* for the various switch settings. If the lamp lights, record a 1, otherwise record a 0. Note that the lamp lights when either or both of the switches are in the 1 position. The logic demonstrator is acting as an OR gate.

OR Gate

A + B		L
0	0	
0	1	
1	0	
1	0	

The NOR Gate

Move the ends of the clips over one place to the NOR terminals. Place switches *A* and *B* initially in the 0 position. Record the condition of lamp *L*, 1 if on and 0 if off. Now try out the other combinations of the switch settings, recording each output condition in the table below.

NOR Gate

$\overline{A + B}$		L
0	0	
0	1	
1	0	
1	1	

After completing the table, compare it to the table for the OR gate. The NOR gate is a NOT OR gate; therefore, the entries in the table under *L* will be opposite. That is, for each entry that is a 0 in one table, there will be a 1 in the other table, and vice versa.

The AND Gate

Move the clip leads over to the right one position and connect them to the terminals marked AND. Again change the positions of switches *A* and *B*, and record your results in the table below.

AND Gate

A • B		L
0	0	
0	1	
1	0	
1	1	

The AND gate is a *coincidence* gate; the lamp should only light when all of the inputs to the AND gate are in the 1 or *on* state.

The NAND Gate

Move the clip leads to the NAND terminals and complete the following truth table.

NAND Gate

A • B		L
0	0	
0	1	
1	0	
1	1	

The NAND gate is a NOT AND gate, so the lamp should not light when the AND gate would, and vice versa. As a result, the only time that the lamp *does not* light with the NAND gate is when all of the inputs are in the 1 or *on* state.

The Exclusive-OR Gate

The exclusive-OR is a complex logic function in that it requires more than one or two AND, OR, and NOT gates to construct it. The circuit in Fig. 9-3 shows how the NAND gates in the logic demonstrator are connected to provide an exclusive-OR function. With the aid of the truth table below, try to reason out the operation of this circuit. With the switches set as shown in Fig. 9-3, determine the logic states of the output of each NAND gate, working from left to right, from the switches to the lamp. To keep track of things, mark down the output logic states on the figure as you work. Then imagine that the switches moved to different positions and repeat your work. (Observe that switch B is a ganged switch in which both switch contacts move together to simultaneously create a 1 and a 0.)

Exclusive-OR Gate

A ⊕ B		L
0	0	
0	1	
1	0	
1	1	

The exclusive-OR circuit shown in Fig. 9-3 is not the simplest that can be constructed using the four NAND gates in the 7400 IC.

242

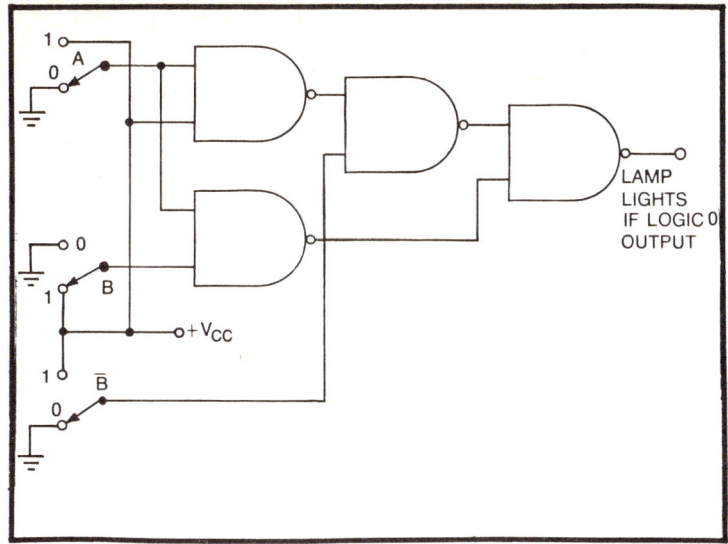

Fig. 9-3. The exclusive-OR gate is a complex gate function requiring several simple gates to construct it.

The logic demonstrator is designed to simulate the operation of many different circuits so some compromise in efficiency must be made to obtain the necessary flexibility. The simplest NAND connection for the exclusive-OR is given in Fig. 9-4A and assumes that both the input logic states and their complements are available; that is A and $\overline{A}$, B and $\overline{B}$.

Alternatively, the complements of A and B can be generated as shown in Fig. 9-4B. Note, however, that this technique requires the use of five NAND gates instead of three. In essence, by connecting one input of each of the first two NAND gates to the positive terminal of the power supply ($+V_{cc}$), a logic-1 input is obtained and these NAND gates function as NOT gates (inverters) to generate the complements of A and B.

While it is possible to construct exclusive-OR gates from either NAND or NOR gates, it is also possible to obtain exclusive-OR gates in a regular logic package. In Chapter 7, for example, the 7486 IC was presented. This is a quad 2-input exclusive-OR gate having inputs and outputs similar to the 7400 quad 2-input NAND. After having seen how a single exclusive-OR gate is constructed, you should be able to appreciate how much circuitry is contained in the 7486 that contains four such gates. Each exclusive-OR gate in the 7486 performs the logic function $A \oplus B = A \cdot \overline{B} + \overline{A} \cdot B$.

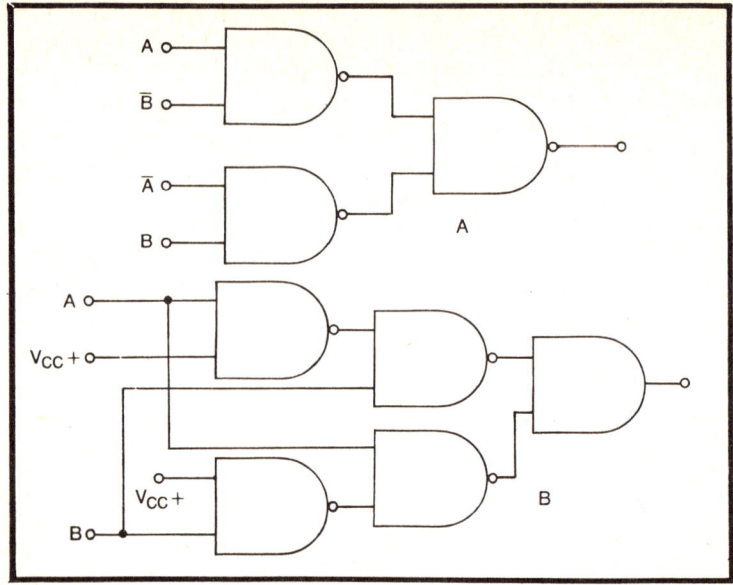

Fig. 9-4. Minimal logic system for exclusive-OR is shown in (A) to produce a low output that will light the lamp in the logic demonstrator. The circuit in (B) is needed to create the logic complements of the A and B signal inputs.

The Exclusive-NOR Gate

By moving the clip leads of the logic demonstrator to the right-most position, the circuit functions as an exclusive-NOR gate. In many textbooks, this function is referred to as the exclusive-AND or the *equality* gate, for the lamp should only

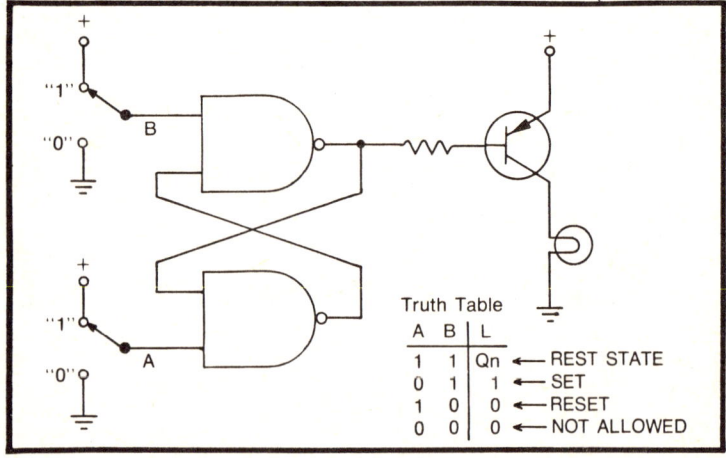

Fig. 9-5. NAND gate flip-flop circuit.

light when the *A* and *B* inputs are equal, both 0 or both 1. In terms of circuitry, however, the only difference between an exclusive-OR and an exclusive-AND is that the outputs are opposite each other—and we know that this can be accomplished by simply adding a NOT gate to the output of the exclusive-OR gate. Thus the exclusive-NOR gate and the exclusive-AND gate perform exactly the same logic function. (It would also be true that the exclusive-OR and the exclusive-NAND are equivalent.)

The logic demonstrator uses still another approach to create the exclusive-NOR function—it simply reverses the connections made to the *B* and $\overline{B}$ sections of the switch. Thus, the lamp lights now when the switches are both 0 or both 1. You can verify this operation yourself using the logic demonstrator and the truth table below.

Exclusive-NOR Gate

A ⊕ B		L
0	0	
0	1	
1	0	
1	1	

There is also a special logic symbol that is sometimes used to represent the exclusive-AND or equality function—a circle with a dot in the middle of it. Consequently, the following logic expressions may be used:

$$\overline{A \oplus B} = A \odot B = A \bullet B + \overline{A} \bullet \overline{B}$$

$$A \oplus B = \overline{A \odot B} = A \bullet \overline{B} + \overline{A} \bullet B$$

NAND FLIP-FLOP

This experiment uses the same 7400 IC that was used in the logic demonstrator in the previous section. You can therefore make use of the same parts to experiment with the NAND flip-flop presented in this section.

The basic building block in TTL is the NAND gate. Similarly, the basic building block in all flip-flops is the RS flip-flop—the simplest of all. Figure 9-5 illustrates how two of the four NAND gates in a 7400 IC can be cross-connected to make an RS flip-flop. If you were to look at the logic diagram of a more complex flip-flop, you would see these same cross-connected gates in the middle, surrounded by other logic gates which merely serve to control their operation.

The flip-flop is remarkable in that it has a "memory." The flip-flop is able to remember which input (A or B) was last in the 0 or *low* logic state. And it continues to remember even after the input has been returned to the 1 or *high* logic state.

The NAND gate RS flip-flop has a "rest" position in which both of its inputs are held in the 1 or high logic state. In the circuit of Fig. 9-5, this is accomplished by setting the A and B switches to the 1 state as shown. The symbol Q_n in the truth table means that the lamp may or may not be on, depending upon whether the flip-flop is "remembering" a 1 or a 0. The lamp will be lit if the flip-flop is storing a logic-1 signal.

If the lamp is not lit, you can "set" the flip-flop by momentarily placing switch A in the 0 position. Note that when the switch is placed in the 0 position the lamp lights immediately. The lamp will then stay lit when switch A is returned to the 1 position. The lamp can be extinguished by "resetting" the flip-flop by momentarily placing switch B in the 0 position. Again, the light will go off immediately when switch B is changed, and the light will stay off when switch B is returned to the 1 position.

A characteristic of all RS flip-flops is that there are only three permissible input combinations. The truth table in Fig. 9-5 indicates that the combination $A = 0$, $B = 0$ is not allowed. This does not mean that the flip-flop will be damaged, but rather that there is no way for the flip-flop to remember that *both* inputs were in the 0 state at the same time. The NAND RS flip-flop will only remember which input was the *last* one to be in the 0 state. (JK flip-flops are unique in that they will toggle—change states—when their two inputs assume this forbidden combination; this useful feature makes it possible for the JK flip-flop to count as well as to store information.)

As an additional project, you might construct an RS flip-flop using NOR gates. You can use a 7402 IC containing four 2-input NOR gates arranged in the same pin configuration as the 7400 IC. The NOR gates would be wired together in the same way as the NAND gates in this experiment. The operation of the flip-flop would be different, however, since the rest position of a NOR gate flip-flop is $A = 0$, $B = 0$. The flip-flop is set and reset by momentarily placing the input switch in the high or 1 position. The input combination $A = 1$, $B = 1$ is not allowed in the NOR gate flip-flop.

MULTIVIBRATOR

The circuit in Fig. 9-6 is an astable multivibrator constructed from the NAND gates in a 7400 IC. The circuit could

246

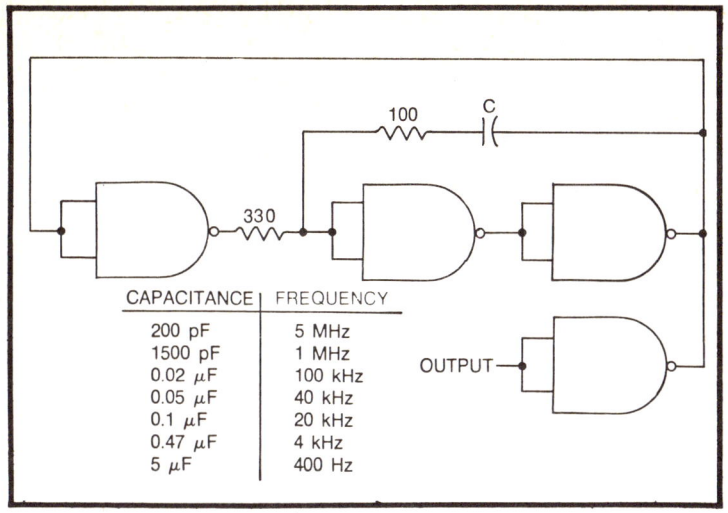

Fig. 9-6. Multivibrator using NAND gates.

also be constructed using the inverter gates in a 7404 package. The NAND gates are made to behave like inverters by connecting their two inputs together as shown in the figure.

The circuit is simple and requires only two resistors and a capacitor. The frequency of the output pulses is varied by changing the value of the capacitor used. The table in the figure provides approximate frequencies obtained with different capacitor values. The output waveform is not a true square wave; the high or 1 state output level does not last quite as long as the low or 0 state output. For audio frequencies, the output of the circuit can be connected to a speaker so that you can hear the output tone. You will obtain a louder output if you connect one terminal of the speaker to the positive battery terminal rather than to ground.

EXPERIMENTER'S POWER SUPPLY

A quick way for the experimenter to make a good but relatively inexpensive power supply is to use the circuits shown in Fig. 9-7. Filament transformers are readily available in 6.3V and 12.6V ratings. And the LM309K voltage regulator is now found in most electronics hobby stores.

For a quick 5V power supply suitable for powering TTL the circuit shown in Fig. 9-7A is ideal. The LM309K regulator provides a very stable 5V output and has internal protection features so that it cannot be damaged by accidental short

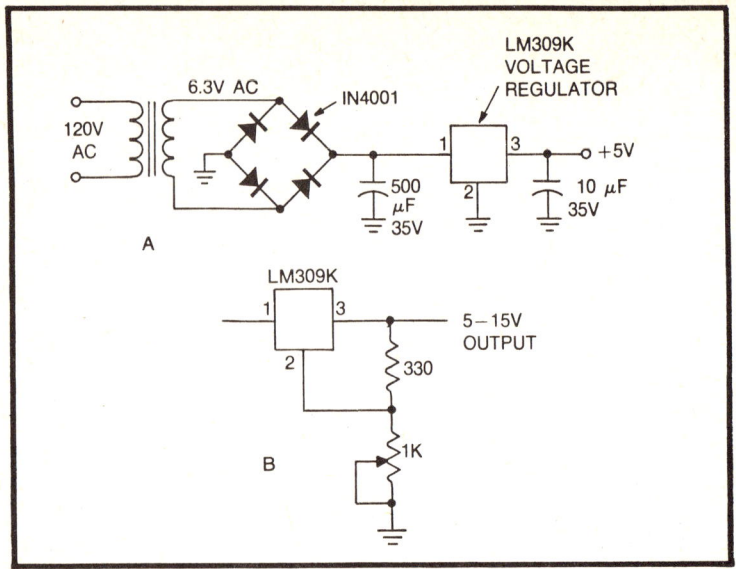

Fig. 9-7. Power supply in (A) produces regulated 5V output that is ideal for powering logic circuits. Optional arrangement in (B) when combined with 12.6V filament transformer can produce variable output for powering a wide variety of circuits.

circuits. The type 1N4001 series of rectifier diodes is also carried by most stores, and there are many replacement types as well. With a 500 μF input filter capacitor, the power supply can provide up to about 100 mA of load current to power your circuits. More current can be obtained by using a larger filter capacitor on the input; the LM309K can supply up to 1.5A if it is mounted on a heat sink.

For higher output voltages, the circuit in Fig. 9-7B can be used. A 12.6V filament transformer is also required. A variable output voltage between 5V and 15V can be obtained by varying the potentiometer setting. The same capacitors can be used as in the 5V circuit. Since a higher input voltage is now available from the 12.6V transformer, the regulator is able to supply several hundred milliamperes in the 5—12V output range with the same 500 μF filter capacitor. Of course, the LM309K tends to heat up more at lower output voltages due to the larger voltage drop between its input and output terminals.

LINEAR AMPLIFIER USING LOGIC GATE

This project demonstrates the versatility of simple IC gates. They can be used not only to synthesize almost any logic

248

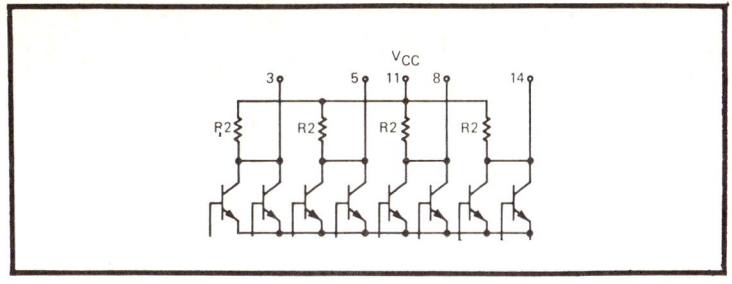

Fig. 9-8. Internal circuitry of HEP 570 quad 2-input NOR gate. (Courtesy Motorola HEP Semiconductors)

function but also can be used as linear amplifiers. The IC used in this circuit is a quad 2-input NOR gate of the RTL family. The internal circuitry of the IC is shown in Fig. 9-8. Note that in connecting together a gate's inputs (pins 1 and 2), for example, we obtain a common-emitter stage consisting of two transistors in parallel. Here, two such stages are formed and connected in cascade.

The result is the high-gain class-A amplifier shown in Fig. 9-9. This circuit is suitable for use as a microphone preamplifier or a low-level audio amplifier. Although a 45Ω speaker is called for, the amplifier may be matched to higher impedance loads by a stepup transformer as shown.

CODE PRACTICE OSCILLATOR

This project, like the previous one, uses a basic gate as an amplifier. Of course, the capability for oscillation is inherent in

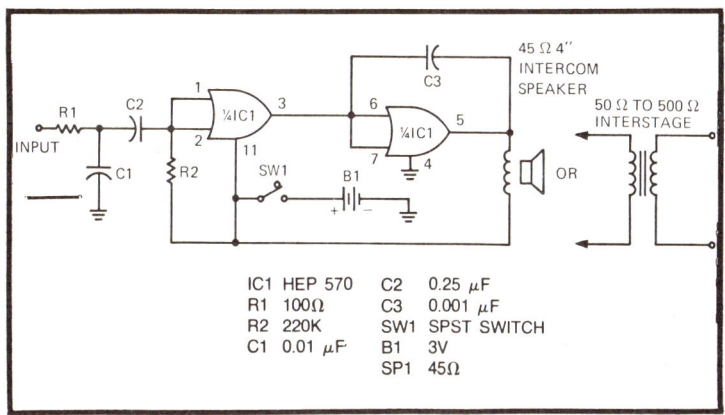

IC1	HEP 570	C2	0.25 μF
R1	100Ω	C3	0.001 μF
R2	220K	SW1	SPST SWITCH
C1	0.01 μF	B1	3V
		SP1	45Ω

Fig. 9-9. A logic gate often makes a good linear amplifier. (Courtesy Motorola HEP Semiconductors)

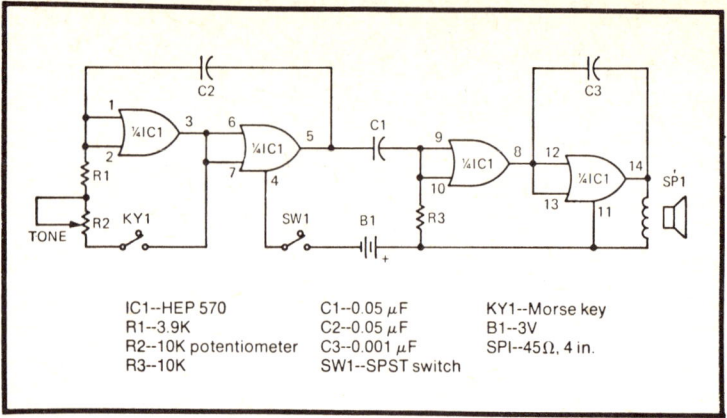

Fig. 9-10. Code practice oscillator. (Courtesy Motorola HEP Semiconductors)

any amplifier. In this case, each gate amplifier stage in Fig. 9-10 shifts the phase of its input signal by 180 degrees; i.e., inverts it. Two such stages in cascade produce a 360-degree phase shift, producing a signal that is in phase with the input of the first stage. By feeding back a signal from the input of the second stage of amplification to the input of the first stage via C2 in the accompanying circuit, we establish the regeneration required to produce oscillation. The circuit shown is simple to construct and

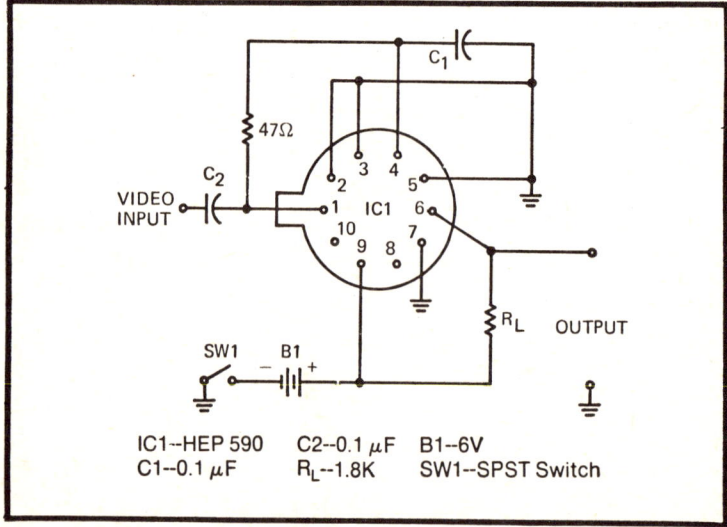

Fig. 9-11. Wide-band video amplifier. (Courtesy Motorola HEP Semiconductors)

produces a loud audio note suitable for use in practicing Morse code for an FCC examination.

VIDEO AMPLIFIER

This project demonstrates one of the most common uses of linear ICs, as a video amplifier. The amplifier in Fig. 9-11 has a bandwidth adequate for handling the composite video signal for color TV. It has 25 dB of gain up to 20 MHz, and 10 dB of gain to 160 MHz. No provisions for tuning are required or provided since it is an untuned wide-band amplifier.

10 MHz OSCILLATOR/TRANSMITTER

The circuit in Fig. 9-12 is a variable frequency oscillator that can be tuned over the 5—10 MHz range, using the parts values shown. The oscillator features a sine-wave output over the full frequency range. The tuning frequency of the oscillator can be scaled down to lower frequencies by increasing the values of C4-T1, the tank circuit. *Warning*: Unless you have an amateur license, it is illegal to use this circuit to transmit over long distances. Therefore, keep the antenna very short to avoid significant radiation beyond the boundaries of your own property.

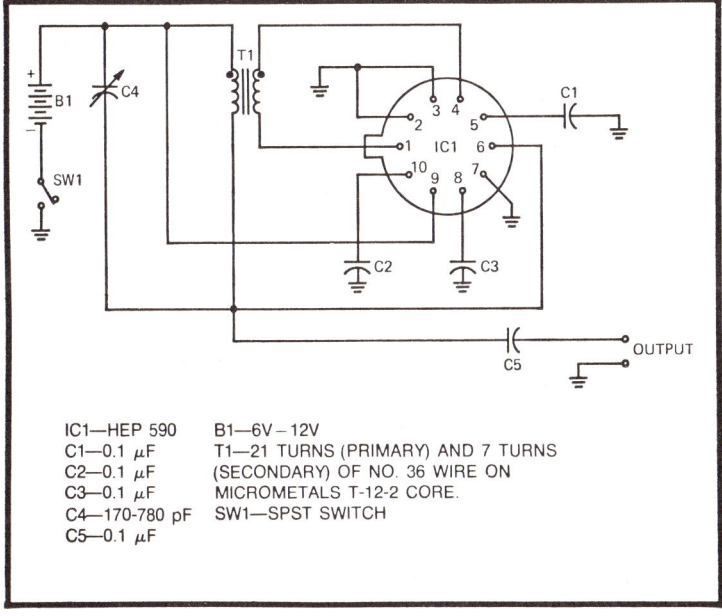

IC1—HEP 590	B1—6V – 12V
C1—0.1 μF	T1—21 TURNS (PRIMARY) AND 7 TURNS
C2—0.1 μF	(SECONDARY) OF NO. 36 WIRE ON
C3—0.1 μF	MICROMETALS T-12-2 CORE.
C4—170-780 pF	SW1—SPST SWITCH
C5—0.1 μF	

Fig. 9-12. 10 MHz oscillator. (Courtesy Motorola HEP Semiconductors)

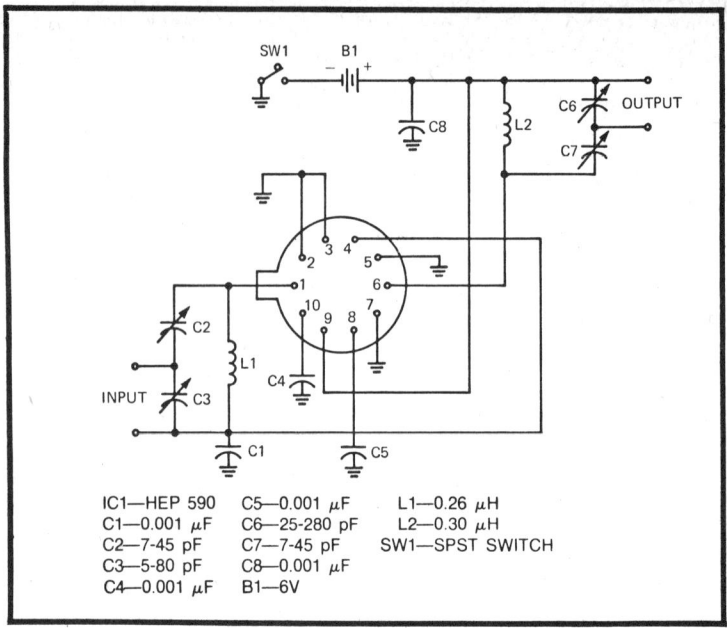

IC1—HEP 590	C5—0.001 μF	L1—0.26 μH
C1—0.001 μF	C6—25-280 pF	L2—0.30 μH
C2—7-45 pF	C7—7-45 pF	SW1—SPST SWITCH
C3—5-80 pF	C8—0.001 μF	
C4—0.001 μF	B1—6V	

Fig. 9-13. 50 MHz preamplifier for amateur radio reception. (Courtesy Motorola HEP Semiconductors)

50 MHz PREAMPLIFIER

This is an amateur radio project utilizing the Motorola HEP 590 linear IC. The HEP 590 is an RF/IF amplifier featuring high gain, extremely low internal feedback, and a wide AGC range. The circuit is a common-emitter, common-base pair, cascade connected, with an AGC transistor and associated biasing circuitry.

Constructing the circuit in Fig. 9-13 will acquaint you with VHF techniques. Observe the following points in VHF work:

- Keep all wiring, including that in bypass circuits, as short and direct as possible.
- When decoupling is required, try a ferrite bead. The bead, which acts as an RF choke, requires no connections but is simply slipped over the interconnecting wiring.
- Use silver-mica or ceramic capacitors of a few picofarads to tune out lead inductances in the input and output circuits.
- Ceramic feedthrough capacitors are useful for bypassing or coupling. Do not use excessive heat when

252

soldering leads to the center hole of the capacitor, as this may destroy the device.

- At VHF, all resistors become capacitive, but quarter-watt resistors less so than higher wattage types.
- Use single-strand wire or copper strap to reduce RF losses.
- The chassis itself may become a resonant cavity at VHF. Watch out for "hot spots" or RF current paths caused by this condition and relocate bypass points if required.

The circuit to be constructed has 30 dB of signal gain with a 0.6 MHz bandwidth. It also has provisions for input and output impedance matching. Grid-dip the circuit to 50.25 MHz. Positive voltage AGC may be added to pin 5 of the IC, but decouple pin 5 to ground using a 0.001 μF capacitor.

ELECTRONIC SIREN

Build the siren circuit shown in the Fig. 9-14. The circuit is useful for a burglar alarm, an automotive theft or tamper alarm, or a toy siren for a bicycle. By slight changes in the setting of the potentiometer, a wide variety of siren sounds can be obtained. *Note*: Do not build the amplifier section of the circuit (shown below dashed line) if the siren is to be used with an external power amplifier.

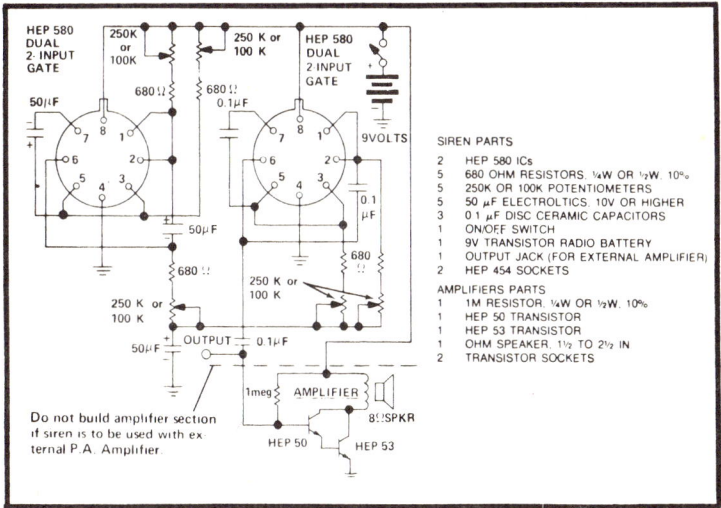

Fig. 9-14. Electronic siren with optional audio amplifier. (Courtesy Motorola HEP Semiconductors)

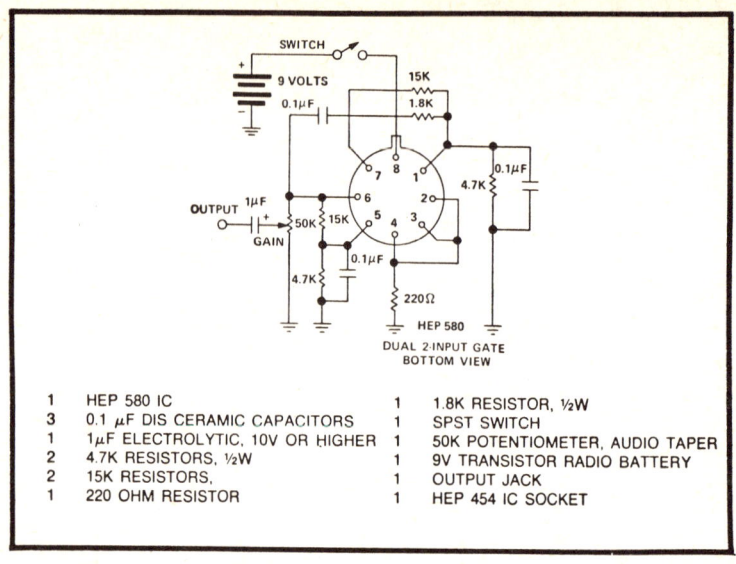

1	HEP 580 IC	1	1.8K RESISTOR, ½W	
3	0.1 μF DIS CERAMIC CAPACITORS	1	SPST SWITCH	
1	1μF ELECTROLYTIC, 10V OR HIGHER	1	50K POTENTIOMETER, AUDIO TAPER	
2	4.7K RESISTORS, ½W	1	9V TRANSISTOR RADIO BATTERY	
2	15K RESISTORS,	1	OUTPUT JACK	
1	220 OHM RESISTOR	1	HEP 454 IC SOCKET	

Fig. 9-15. Audio signal generator. (Courtesy Motorola HEP Semiconductors)

AUDIO SIGNAL GENERATOR

A fixed-frequency sine wave having a 2V peak-to-peak output can be obtained from the audio generator in Fig. 9-15. For the values indicated, the output frequency is approximately 1 kHz. The two 4.7K resistors and two 0.1 μF bridging capacitors determine the frequency of oscillation. Although small variations in the resistance values are permissible, the most effective way in which to change the operating frequency is to substitute another pair of capacitors for the 0.1 μF units illustrated.

FOUR-CHANNEL AUDIO MIXER

A typical audio preamplifier or tuner will provide an output of about 0.25V peak-to-peak. A four-channel (quadraphonic) input to the circuit in Fig. 9-16 can be mixed and amplified to provide an output of about 4.5V peak-to-peak. For individual level control of each input channel, a 10K audio-taper potentiometer and a 10 μF, 10V capacitor can be added to each input as illustrated by the dashed lines in the figure. With a 9V power source, current drain is 3 mA.

WIDE-BAND PREAMPLIFIER

The simple preamplifier circuit in Fig. 9-17 may be used to amplify low-level signals before being applied to an oscilloscope

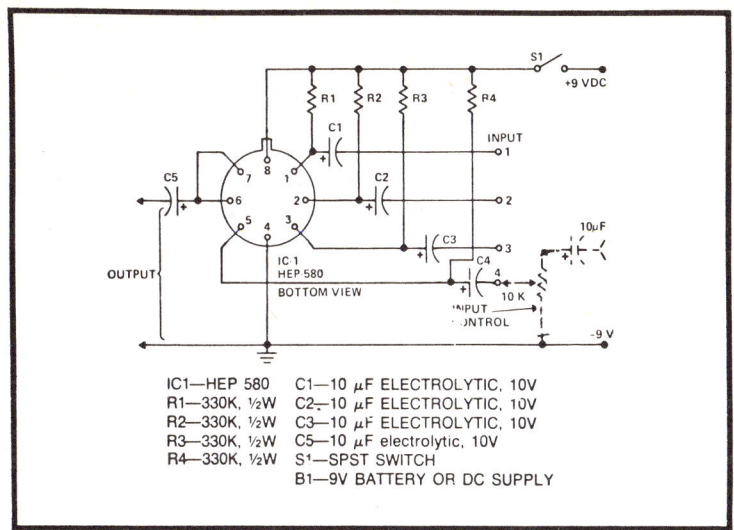

IC1—HEP 580   C1—10 μF ELECTROLYTIC, 10V
R1—330K, ½W   C2—10 μF ELECTROLYTIC, 10V
R2—330K, ½W   C3—10 μF ELECTROLYTIC, 10V
R3—330K, ½W   C5—10 μF electrolytic, 10V
R4—330K, ½W   S¹—SPST SWITCH
                 B1—9V BATTERY OR DC SUPPLY

Fig. 9-16. Four-channel audio mixer is good for microphone mixing and for quadraphonic-to-mono mixing. (Courtesy Motorola HEP Semiconductors)

or AC voltmeter. It is also a useful preamplifier when used with microphones, phonographs, and hearing aids. You'll probably think of dozens of other uses.

The preamplifier can provide lots of gain. With a 9V power supply, the preamplifier can provide a 1V output with as little as 0.01V input signal. Frequency response is 10 Hz to 1 MHz. It's ideal for battery operation since it draws only 0.2 mA.

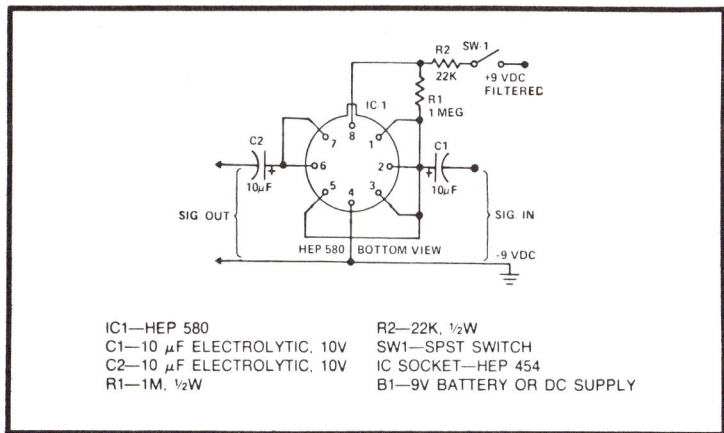

IC1—HEP 580   R2—22K, ½W
C1—10 μF ELECTROLYTIC, 10V   SW1—SPST SWITCH
C2—10 μF ELECTROLYTIC, 10V   IC SOCKET—HEP 454
R1—1M, ½W   B1—9V BATTERY OR DC SUPPLY

Fig. 9-17. Wide-band preamplifier for good signal amplification up to 1 MHz. (Courtesy Motorola HEP Semiconductors)

255

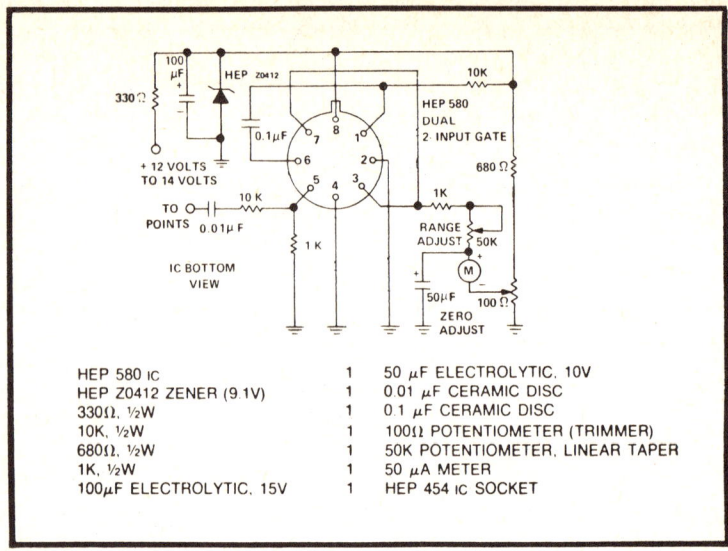

HEP 580 IC	1	50 µF ELECTROLYTIC, 10V
HEP Z0412 ZENER (9.1V)	1	0.01 µF CERAMIC DISC
330Ω, ½W	1	0.1 µF CERAMIC DISC
10K, ½W	1	100Ω POTENTIOMETER (TRIMMER)
680Ω, ½W	1	50K POTENTIOMETER, LINEAR TAPER
1K, ½W	1	50 µA METER
100µF ELECTROLYTIC, 15V	1	HEP 454 IC SOCKET

Fig. 9-18. Versatile IC tachometer is easy to build and easy to calibrate. (Courtesy Motorola HEP Semiconductors)

BUILD AN IC TACHOMETER

This versatile circuit (Fig. 9-18) can be used with automobiles, boats, and motorcycles. The rpm range and

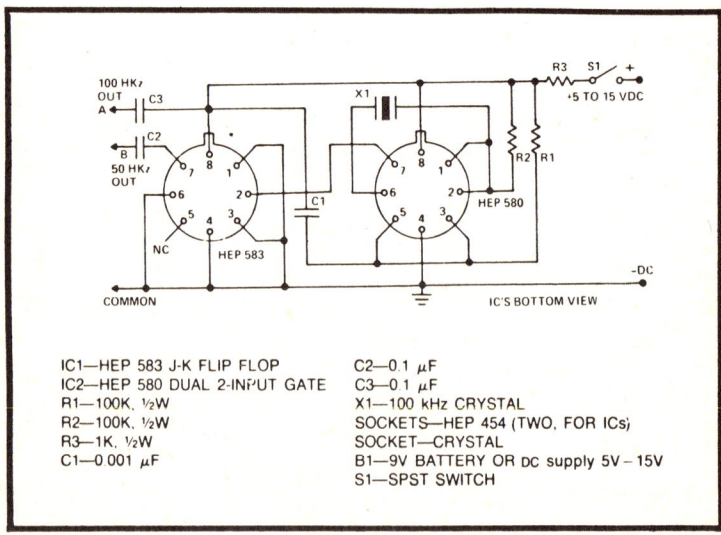

IC1—HEP 583 J-K FLIP FLOP	C2—0.1 µF
IC2—HEP 580 DUAL 2-INPUT GATE	C3—0.1 µF
R1—100K, ½W	X1—100 kHz CRYSTAL
R2—100K, ½W	SOCKETS—HEP 454 (TWO, FOR ICs)
R3—1K, ½W	SOCKET—CRYSTAL
C1—0.001 µF	B1—9V BATTERY OR DC supply 5V – 15V
	S1—SPST SWITCH

Fig. 9-19. Precision 100 kHz oscillator. (Courtesy Motorola HEP Semiconductors)

number of cylinders can be compensated for with just one adjustment.

Connect the circuit ground to the vehicle ground and the +12V input to the positive terminal of the vehicle battery. With the engine turned *off* but the ignition switch turned *on*, connect the tachometer to the ignition points. Zero the meter with the ZERO ADJUST potentiometer. Then with the RANGE ADJUST potentiometer, calibrate the tachometer, using a high-quality tachometer from a garage as a standard of comparison. The range maximum can be whatever you desire; calibrate the rest of the meter scale accordingly.

100 kHz PRECISION FREQUENCY STANDARD

Using a 100 kHz crystal, the square-wave generator circuit in Fig. 9-19 provides accurate crystal-controlled outputs at both 50 and 100 kHz. These square-wave signals are rich in harmonics, so they are especially useful in producing accurate frequency markers in communications receivers.

When connected to a 9V battery or other power supply, the generator draws just 5.5 mA. It provides approximately 1V peak-to-peak at signal outputs A and B.

Index

Index

263